Repeat Amino Acid Sequences in Proteins

Biochemical and Medical Consequences

Editors

Vladimir N. Uversky

Department of Molecular Medicine
University of South Florida
Tampa, USA

Munishwar Nath Gupta

Former Emeritus Professor
Department of Biochemical Engineering and Biotechnology
Indian Institute of Technology Delhi
New Delhi, India

CRC Press
Taylor & Francis Group
Boca Raton London New York

CRC Press is an imprint of the
Taylor & Francis Group, an **informa** business
A SCIENCE PUBLISHERS BOOK

Cover credit: Image taken from Chapter 1 of this book. Reproduced by kind courtesy of authors of the chapter.

First edition published 2026
by CRC Press
2385 NW Executive Center Drive, Suite 320, Boca Raton FL 33431

and by CRC Press
4 Park Square, Milton Park, Abingdon, Oxon, OX14 4RN

CRC Press is an imprint of Taylor & Francis Group, LLC

Library of Congress Cataloging-in-Publication Data (applied for)

ISBN: 978-1-032-82445-1 (hbk)
ISBN: 978-1-032-83562-4 (pbk)
ISBN: 978-1-003-50996-7 (ebk)

DOI: 10.1201/9781003509967

Typeset in Palatino
by Prime Publishing Services

Preface

The thought of putting together the present book came up after we had just co-authored an extensive review on Arginine (Gupta and Uversky 2024a) which in turn led us to speculate that many of the properties of Arginine may in fact be due to its tendency to form clusters and repeat sequences (Gupta and Uversky 2024b). In the present book, we examine the biological importance of repeat sequences of other amino acids as well.

The first chapter presented by Yuzo Fujino and Yoshitaka Nagai discusses a general phenomenon of repeat expansion diseases associated with the expanded short tandem repeats (STRs). There are at least 50 such repeat expansion diseases, which originate from the STR expansions happening in coding or noncoding regions of the corresponding genes. In addition to the polyQ maladies, coding repeat expansion diseases include polyalaninine (polyA) and polyglycine (polyG) diseases, originating from the expansion of the CAG, GCN, and GGC repeat sequences, respectively. Noncoding repeat expansion diseases are caused by the expansion of trinucleotide, tetranucleotide, pentanucleotide, hexanucleotide, or even dodecanucleotide sequences. The corresponding sequences can be translated via a noncanonical form of translation, repeat-associated non-AUG (RAN) translation, which is characterized by the AUG codon-independent and repeat sequence-dependent translation initiation, and which can happen from all reading frames.

This chapter also delves into how these expansions correlate with intrinsic disorder, which leads them also to play important roles in the dynamics of the reversible formation of the membrane less organelles in cells via liquid-liquid phase separation.

While the chapters in this book can be read in any order and independently, it will be helpful to read this first chapter as an overview of the fairly wide footprint of the repeat sequences in biology as well as medical sciences.

The second chapter by Raffaella Parlato and Patrick C.A. van der Wel is dedicated to the introduction of a class of protein aggregation diseases associated with the expansion of the repetitive polyglutamine (polyQ) segment (also known as polyQ or CAG-repeat expansion disorders or polyQ-related syndromes). This chapter is mostly focused on the polyQ expansion-driven Huntington's disease. Although nine neurodegenerative diseases are associated with expansion of polyQ region of specific proteins, the most well-known example is Huntington's disease, which arises as a consequence of the polyQ repeat expansion in the huntingtin (HTT) protein above a specific threshold (35 glutamines or 35 CAG codons in the *HTT* gene).

Biochemists sometimes tend to overlook that a large number of proteins have hydrophobic clusters on their surfaces. The third chapter by Fabio C.L. Almeida, Ramon Pinheiro-Aguiar and Ariana Azevedo Vasconcelos though deals with a special class of proteins, which do not have a hydrophobic core but have stable 3D structures secured by multiple disulfide (S–S) bonds and contain surface hydrophobic clusters important for their function. These proteins represent an interesting case of intrinsically disordered proteins that are capable of gaining stable structures as a result of the oxidative folding. Readers will note that the intrinsic disorder runs as a common thread in case of cellular functions of proteins with repeat sequences, which are discussed in various chapters.

The roles of various classes of amino acid clusters in protein-protein interactions is the theme of the fourth chapter. Here, Juan Martín Hernández Castillo, Albros Hermes Poot and Gabriel Del Río introduce Residue Cluster Classes (RCC), which are maximally connected clusters of amino acids found in protein structures. RCC serves as a useful means for the efficient classification of protein structure and prediction of protein function and protein-protein interactions.

The fifth chapter by Esteban Lanzarotti, Lucas Defelipe, Marcelo Marti and Adrian Turjanski specifically discusses aromatic clusters and also covers their importance in protein-drug complex formations. The importance of aromatic residues for structure and function of proteins is difficult to overestimate. It was shown that protein data bank (PDB) virtually does not contain "non-smelly" proteins, i.e., proteins that do not contain cysteine and aromatic residues (CFYWH-depleted), indicating that that unique 3D structures of monomeric globular proteins are crucially dependent on the presence of cysteine and aromatic residues. Overall, three major roles of aromatic clusters have been outlined, where intraprotein (IP) aromatic clusters are formed by interactions between residues of the same protein chain and thereby are only involved in protein folding; protein-drug (PD) aromatic clusters are present at the sites where drug-like ligands are bound; and protein-

protein (PP) aromatic clusters are found at the protein-protein interaction interfaces.

Cell-penetrating peptides represent another set of molecules, which have repeat amino acid sequences. These peptides (both of natural origin and artificially designed) can cross cell membranes without compromising their structural or functional integrity and thereby serve as promising means for the intracellular delivery of various compounds ranging from small molecules to peptides, proteins, and nucleic acids. The sixth chapter by Ngoc Hai Trieu Phong, Mohammed Kamal, Duy Binh Tran and Tohru Yamada covers the ever-increasing importance of such cell-penetrating peptides (many of which contain repeat amino acid motifs), particularly in the context of the design of novel drug-delivery systems with high efficiency and controllable specificity.

Elastin-like peptides gained attention initially as a fusion tag for convenient purification of recombinant proteins. These now constitute a hugely versatile class of molecules, which straddle a large number of areas in biotechnology and biomedical sciences. The seventh chapter by Takeru Nose and Keitaro Suyama is a good overview of these peptides and their importance in both biology and biotechnology.

All textbooks of biochemistry, enzymology, and protein science typically identify aromatic amino acids as chromophores responsible for the intrinsic fluorescence of proteins. The last and eighth chapter by Vishnu N. Vijayan, Karthika Kannan and Sushabhan Sadhukhan describe the fluorescence shown by some non-aromatic amino acids, when these occur as repeat sequences. Fluorescence-based tools and techniques have now become immensely powerful in the study of biological molecules and processes. The applications of the fluorescence shown by the non-aromatic amino acids which are described in this interesting chapter may turn out to be the proverbial tip of the ice berg.

We hope that the book will be found useful by diverse kinds of readers who are interested in the involvement of peptides and proteins in biology, biotechnology, drug delivery designs and sensors and many other areas.

We thank all the contributing authors for their efforts.

Editors
Munishwar Nath Gupta
Vladimir N. Uversky

REFERENCES

Gupta, M.N. and V.N. Uversky. 2024a. Biological importance of arginine: A comprehensive review of the roles in structure, disorder, and functionality of peptides and proteins. Int. J. Biol. Macromol 257(Pt. 1): 128646. doi: 10.1016/j.ijbiomac.2023.128646.

Gupta, M.N. and V.N. Uversky. 2024b. Reexamining the diverse functions of arginine in biochemistry. Biochem Biophys Res Commun 705: 149731. doi: 10.1016/j.bbrc.2024.149731.

Contents

Chapter **1**

Pathogenic Role of Repeat Polypeptides in Repeat Expansion Diseases

Yuzo Fujino*[1] and Yoshitaka Nagai*[1,2]

[1]Department of Neurology, Kindai University, Faculty of Medicine, Osaka-Sayama, Osaka 589-8511, Japan.

[2]Life Science Research Institute, Kindai University, Osaka-Sayama, Osaka 589-8511, Japan

INTRODUCTION

The human genome contains numerous repetitive sequences, among which simple repetitive sequences consisting of a short repeat motif of 2 to 12 bases are known as Short Tandem Repeats (STRs). STRs have more than one million loci, and constitute as much as 3% of the human genome (Malik et al. 2021). The expansion of STRs can cause monogenic diseases, including various neuromuscular conditions, and are collectively referred to as repeat expansion diseases. Since the identification of the CGG repeat expansion in the *FMR1* gene and the CAG repeat expansion in the androgen receptor gene as the causative mutations for Fragile X Syndrome (FXS) and Spinal and Bulbar Muscular Atrophy (SBMA), respectively, in 1991 (Kremer et al. 1991, La Spada et al. 1991),

*For Correspondence: fujino@koto.kpu-m.ac.jp; yoshi.nagai@med.kindai.ac.jp

more than 50 repeat expansion diseases have been identified so far (Malik et al. 2021, Fujino and Nagai 2022).

Repeat expansion diseases can be classified into two groups, namely, coding and the noncoding repeat expansion diseases. In coding repeat expansion diseases, expanded STRs are located within the coding regions of the mutant genes, and are translated into expanded amino acid repeat sequences within the mutant proteins. In particular, in polyglutamine diseases caused by the expansion of CAG repeat sequences, the expanded polyglutamine tracts in the mutant proteins are primarily thought to contribute to the pathogenesis. On the other hand, in the noncoding repeat expansion diseases, expanded STRs are located in the noncoding regions, such as introns and untranslated regions, and thus the transcribed mutant RNA containing aberrantly expanded repeat sequences (repeat RNA) has been considered to be the key pathological molecule. However, recent studies have revealed that repeat RNA can also be translated into repeat polypeptides through a noncanonical mechanism known as Repeat-Associated Non-AUG (RAN) translation, even in the absence of an AUG initiation codon (Zu et al. 2011). These RAN-translated repeat polypeptides have indeed been detected in patients' tissues, and are thought to play crucial roles in the pathogenesis of the noncoding repeat expansion diseases (Fujino et al. 2023a). The discovery of RAN translation suggests common pathomechanisms mediated by repeat polypeptides in both the coding and noncoding repeat expansion diseases. Notably, these repeat polypeptides have been found to disrupt physiological Liquid-Liquid Phase Separation (LLPS) of proteins and RNAs, leading to the dysregulation of biological condensates called membraneless organelles (MLOs) (Fujino and Nagai 2022). This chapter outlines the major pathomechanisms underlying repeat expansion diseases, with a particular focus on the roles of repeat polypeptides.

CODING REPEAT EXPANSION DISEASES

The coding repeat expansion diseases can be categorized into the following three subtypes depending on the encoding repeat sequences: polyglutamine diseases, polyalanine diseases, and polyglycine diseases.

The polyglutamine diseases are a group of hereditary neurological diseases caused by expanded CAG repeat sequences that encode polyglutamine. In addition to the aforementioned SBMA, Huntington disease, spinocerebellar ataxia (SCA) types 1, 2, 3, 6, 7, 17, and 51, and dentatorubral pallidoluysian atrophy belong to this disease entity (Takeuchi and Nagai 2017, Tan et al. 2023) (Table 1.1). The polyglutamine diseases share the following clinicogenetic characteristics: (1) the

Table 1.1 Coding repeat expansion diseases

Disease	**Gene**	**Inheritance**	**Repeat motif**	**Repeat length**	
				Normal	**Disease**
Polyglutamine diseases					
Dentatorubral-pallidoluysian atrophy	*ATN1*	AD	CAG	6–36	48–93
Huntington disease	*HTT*	AD	CAG	6–35	36–250
Spinocerebellar ataxia 1	*ATXN1*	AD	CAG	6–39	39–91
Spinocerebellar ataxia 2	*ATXN2*	AD	CAG	14–31	32–200
Spinocerebellar ataxia 3	*ATXN3*	AD	CAG	10–44	53–87
Spinocerebellar ataxia 6	*CACNA1A*	AD	CAG	4–19	20–33
Spinocerebellar ataxia 7	*ATXN7*	AD	CAG	4–35	36–306
Spinocerebellar ataxia 17	*TBP*	AD	CAG	25–44	45–63
Spinocerebellar ataxia 51	*THAP11*	AD	CAG	20–38	45–100
Spinal and bulbar muscular atrophy	*AR*	XR	CAG	9–36	38–62
Polyalanine diseases					
Blepharophimosis, ptosis, and epicanthus inversus syndrome	*FOXL2*	AD	GCN	14	22–24
Cleidocranial dysplasia	*RUNX2*	AD	GCN	17	27
Congenital central hypoventilation syndrome	*PHOX2B*	AD	GCN	20	25–33
Developmental and epileptic encephalopathy 1	*ARX*	XR	GCN	12–16	20–27
Hand-foot-genital syndrome	*HOXA13*	AD	GCN	18	24–27
Hereditary sensory and autonomic neuropathy 8	*PRDM12*	AR	GCN	7–14	18–19
Holoprosencephaly 5	*ZIC2*	AD	GCN	15	25
Oculopharyngeal muscular dystrophy	*PABPN1*	AD/AR	GCN	10	12–17
Synpolydactyly 1	*HOXD13*	AD	GCN	10	22–29
X-linked mental retardation with isolated growth hormone deficiency	*SOX3*	XR	GCN	15	26
Polyglycine disease					
Spinocerebellar ataxia 4	*ZFHX3*	AD	GGC	14–26	42–74

Abbreviations: AD, autosomal dominant; AR, autosomal recessive; XR, X-linked recessive

threshold number of expanded CAG repeats is similar among the diseases (approximately 35–40 repeats); (2) the number of CAG repeats correlates with disease severity, and inversely correlates with the age of onset; and

(3) they are inherited in an autosomal-dominant manner, except for SBMA, which shows X-linked inheritance. In addition, there is no homology nor functional similarity among the causative genes. These characteristics strongly indicate that expanded polyglutamine tracts within the mutant proteins themselves primarily contribute to neurodegeneration through gain-of-toxic function mechanisms, irrespective of the functions of the causative genes (Takeuchi and Nagai 2017). Indeed, it has been shown that expansion of polyglutamine tracts induces misfolding and a β-sheet conformational transition of the mutant proteins, resulting in the formation of amyloid-like fibrillar aggregates and their accumulation as inclusions in the brain, leading to neurodegeneration (Nagai et al. 2007) (Figure 1.1). Importantly, the β-sheet conformational transition of the polyglutamine protein monomer, not its aggregation into amyloid-like fibrils, triggers cytotoxicity (Nagai et al. 2007). Based on these findings, screening for small molecules has identified arginine, one of chemical chaperones, that stabilizes the conformation of polyglutamine proteins, inhibits their aggregation, and exerts therapeutic effects on animal models of the polyglutamine diseases (Minakawa et al. 2020). A clinical trial of arginine for patients with SCA6 has been conducted in Japan, with promising results (Ishihara et al. 2024).

Polyalanine diseases, on the other hand, comprise nine congenital diseases, including neurodevelopmental disorders and polydactyly, as well as late-onset oculopharyngeal muscular dystrophy, all of which are caused by the expansion of GCN repeat sequences encoding polyalanine (Albrecht and Mundlos 2005, Chen et al. 2015) (Table 1.1). In congenital polyalanine diseases, a loss-of-function of the causative genes is thought to play a major role in the pathogenesis, since these genes encode transcription factors crucial for development, and missense/nonsense/frameshift mutations in the same genes have also been found to reproduce similar phenotypes (Albrecht and Mundlos 2005). However, expanded polyalanine tracts themselves may also contribute to the pathomechanisms to some extent, because the number of expanded GCN repeats correlates with disease severity (Goodman et al. 1997, Kato et al. 2004, Matera et al. 2004). Recent studies have shown that mutant HOXD13 and other transcription factors containing expanded polyalanine tracts perturb the physiological composition of transcriptional condensates, which are nuclear MLOs composed of these transcription factors and coactivators, leading to the dysregulation of transcriptional gene expression program (Basu et al. 2020).

Polyglycine diseases are another group of repeat expansion diseases caused by the expansion of GGC repeat sequences encoding polyglycine. This disease group encompasses both the coding and noncoding repeat expansion diseases, depending on the location of the mutations within the genes. SCA4 (Chen et al. 2024, Figueroa et al. 2024, Paucar et al.

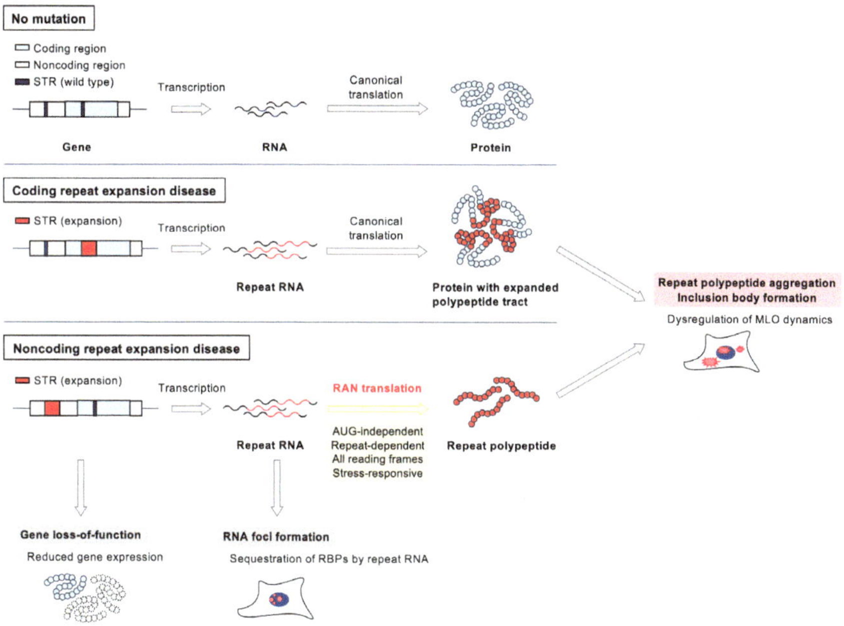

Figure 1.1 Overview of the pathomechanisms underlying repeat expansion diseases. In coding repeat expansion diseases represented by the polyglutamine diseases, the expanded polyglutamine tracts in mutant proteins cause their misfolding and β-sheet conformational transition, resulting in the formation of amyloid-like fibrillar aggregates of the mutant proteins, contributing to neurodegeneration. In noncoding repeat expansion diseases, repeat RNAs are translated into repeat polypeptides through noncanonical RAN translation, which eventually aggregate, contributing to neurodegeneration. In both coding and noncoding repeat expansion diseases, these repeat polypeptides disrupt the physiological LLPS of proteins and RNAs, leading to the dysregulation of MLOs. In addition to the repeat polypeptide-mediated pathomechanisms, RNA foci formation by repeat RNA and the loss-of-function of the causative genes also contribute to neurodegeneration.

Abbreviations: LLPS, liquid-liquid phase separation; MLO, membraneless organelle; STR, short tandem repeat; RAN, repeat-associated non-AUG; RBP, RNA-binding protein

2024, Wallenius et al. 2024) belongs to the former category, while Fragile X Tremor/Ataxia Syndrome (FXTAS) (Todd et al. 2013), Neuronal Intranuclear Inclusion Disease (NIID) (Boivin et al. 2021, Zhong et al. 2021), and oculopharyngodistal myopathy (OPDM) (Yu et al. 2021, Yu et al. 2022) belong to the latter category. SCA4 is caused by expanded GGC repeat sequences in the coding exon of the *ZFHX3* gene, which has been recently identified (Chen et al. 2024, Figueroa et al. 2024, Paucar et al. 2024, Wallenius et al. 2024) (Table 1.1). Patients with SCA4 exhibit characteristic neurological symptoms, such as progressive ataxia, sensory neuropathy, and autonomic dysfunction. Although the detailed pathomechanisms of this disease remain unclear, gain-of-toxic function mechanisms mediated by expanded GGC repeat sequences have been suggested, since the disease shares some characteristics with the polyglutamine diseases; i.e., it is inherited in an autosomal-dominant manner, the number of GGC repeats correlates with the age of onset and the severity of symptoms (Dalski et al. 2024, Figueroa et al. 2024), and aberrant accumulation of ZFHX3 proteins as well as polyglycine proteins in the brain has been observed neuropathologically (Figueroa et al. 2024, Paucar et al. 2024). In contrast, loss-of-function mutations in the *ZFHX3* gene result in syndromic intellectual disability, a phenotype typically not observed in SCA4 (Pérez Baca et al. 2024). Given that polyglycine proteins have also been implicated in the pathogenesis of several noncoding repeat expansion diseases, such as the aforementioned FXTAS, NIID, and OPDM, the possible common polyglycine-mediated gain-of-toxic pathomechanisms between SCA4 and these diseases are attracting substantial attention in this research field.

NONCODING REPEAT EXPANSION DISEASES

The noncoding repeat expansion diseases encompass a diverse group of neuromuscular diseases, including the aforementioned FXS, FXTAS, *C9orf72*-linked amyotrophic lateral sclerosis/frontotemporal dementia (C9-ALS/FTD), SCA types 8, 10, 12, 27B, 31, 36, and 37, NIID, myotonic dystrophy (DM) types 1 and 2, and OPDM types 1–4, etc., (Table 1.2). These diseases are caused by the expansion of trinucleotide, tetranucleotide, pentanucleotide, hexanucleotide, or even dodecanucleotide sequences, with the number of expanded STRs varying widely, ranging from approximately 50 to more than 10,000, with a much greater variability than the coding repeat expansion diseases (Fujino and Nagai 2022).

In some noncoding repeat expansion diseases, such as FXS and Friedreich ataxia, missense mutations in the same causative genes have also been identified, suggesting that loss-of-function mechanisms would contribute to their pathogenesis (Myrick et al. 2014, Bürk 2017) (Figure 1.1). For the other diseases, repeat RNA transcribed from the mutant genes had long been considered to be the key toxic molecule involved in the pathomechanisms. Repeat RNA aberrantly interacts with various RNA-Binding Proteins (RBPs), thereby altering their function, distribution, and stability. In particular, repeat RNA can form aggregates known as RNA foci in affected cells, which sequester the RBPs, leading to a loss of their function and dysregulation of RNA metabolism (Figure 1.1). For example, in DM1, the splicing regulator MBNL1 is sequestered into RNA foci formed by CUG repeat RNA, resulting in its functional loss

Table 1.2 Noncoding repeat expansion diseases

Disease	Gene	Location	Inheritance	Repeat motif	Repeat length	
					Normal	Disease
Baratela-Scott syndrome	*XYLT1*	promoter	AR	CGG	9–20	100–830
C9orf72-linked amyotrophic lateral sclerosis/ frontotemporal dementia	*C9orf72*	5′-UTR intron	AD	GGGGCC	2–23	700–1,600
Cerebellar ataxia, neuropathy, and vestibular areflexia syndrome	*RFC1*	intron	AR	AAGGG	11	400–2,000
Familial adult myoclonic epilepsy 1	*SAMD12*	intron	AD	TTTCA	0	105–3,680*
Familial adult myoclonic epilepsy 2	*STARD7*	intron	AD	TTTCA	0	150–460
Familial adult myoclonic epilepsy 3	*MARCH6*	intron	AD	TTTCA	0	660–2,800*
Familial adult myoclonic epilepsy 4	*YEATS2*	intron	AD	TTTCA	0	1,000–1,600*
Familial adult myoclonic epilepsy 6	*TNRC6A*	intron	AD	TTTCA	0	–1,100
Familial adult myoclonic epilepsy 7	*RAPGEF2*	intron	AD	TTTCA	0	–2,200
Fragile X syndrome	*FMR1*	5′-UTR	XR	CGG	5–40	>200
Fragile X-associated tremor/ataxia syndrome	*FMR1*	5′-UTR	XD	CGG	5–40	55–200
Fragile XE mental retardation	*FMR2*	5′-UTR	XR	CCG	6–35	>200
Friedreich ataxia	*FXN*	intron	AR	GAA	5–33	66–1,300
Fuchs endothelial corneal dystrophy 3	*TCF4*	intron	AD	CTG	<40	>50
Global developmental delay, progressive ataxia, and elevated glutamine	*GLS*	5′-UTR	AR	GCA	5–26	>90
Huntington disease-like 2	*JPH3*	3′-UTR	AD	CTG	6–28	41–58
LRP12-linked amyotrophic lateral sclerosis	*LRP12*	5′-UTR	AD	CGG	2–49	64–100

(**Table 1.2** *Contd...*)

Table 1.2 *Contd...*

Disease	Gene	Location	Inheritance	Repeat motif	Repeat length	
					Normal	Disease
LRP12-linked intertied peripheral neuropathy	*LRP12*	5′-UTR	AD	CGG	2–49	50–152
Myoclonus epilepsy of Unverricht and Lundborg	*CSTB*	promoter	AR	CCCCGC-CCCGCG	2–3	30–125
Myotonic dystrophy 1	*DMPK*	3′-UTR	AD	CTG	5–37	50–> 2,000
Myotonic dystrophy 2	*ZNF9*	intron	AD	CCTG	11–26	75–11,000
Neuronal intranuclear inclusion disease	*NOTCH2NLC*	5′-UTR	AD	CGG	5–39	66–517
Oculopharyngeal myopathy with leukoencephalopathy	*LOC642361/ NUTM2BAS1*	noncoding	AD	CGG	3–16	N/A**
Oculopharyngodistal myopathy 1	*LRP12*	5′-UTR	AD	CGG	2–49	76–630
Oculopharyngodistal myopathy 2	*GIPC*	5′-UTR	AD	CGG	4–29	73–164
Oculopharyngodistal myopathy 3	*NOTCH2NLC*	5′-UTR	AD	CGG	5–39	128–198
Oculopharyngodistal myopathy 4	*RILPL1*	5′-UTR	AD	CGG	9–16	139–197
Spinocerebellar ataxia 8	*ATXN8OS*	3′-UTR	AD	CTG	15–50	71–1,300
Spinocerebellar ataxia 10	*ATXN10*	intron	AD	ATTCT	10–29	280–4,500
Spinocerebellar ataxia 12	*PPP2R2B*	5′-UTR	AD	CAG	7–45	51–78
Spinocerebellar ataxia 27B	*FGF14*	intron	AD	GAA	8–250	250–> 700
Spinocerebellar ataxia 31	*BEAN1*	intron	AD	TGGAA	0	500–760
Spinocerebellar ataxia 36	*NOP56*	intron	AD	GGCCTG	3–14	650–2,500
Spinocerebellar ataxia 37	*DAB1*	5′-UTR	AD	ATTTC	0	31–75

*Total length of TTTCA and TTTTA repeats.
**Difficult to be confirmed due to somatic instability
Abbreviations: AD, autosomal dominant; AR, autosomal recessive; UTR, untranslated region; XD, X-linked dominant; XR, X-linked recessive

and the missplicing of its target genes, including *CLCN1*. The reduced expression of *CLCN1* contributes to myotonia, a hallmark symptom of this disease (Mankodi et al. 2002, Lin et al. 2006).

Moreover, it has surprisingly been revealed that repeat RNA can be translated into repeat polypeptides even in the absence of AUG initiation codons, via an unconventional mechanism known as RAN translation (Zu et al. 2011). RAN-translated repeat polypeptides have been pathologically confirmed in patients with various diseases, and have been reported to be toxic by disrupting multiple biological pathways, including nucleocytoplasmic transport, axonal transport, global translation, and the proteasomal system, in cellular and animal models (Fujino et al. 2023a). Consequently, RAN-translated repeat polypeptides are now recognized as novel key pathological molecules through gain-of-toxic function mechanisms in the pathogenesis of the noncoding repeat expansion diseases, together with toxic repeat RNA and the loss-of-function of the causative genes (Figure 1.1). In the following section, the general characteristics and regulatory mechanisms of RAN translation will be discussed.

RAN TRANSLATION

In 2011, Ranum and colleagues first identified a noncanonical form of translation occurring from the CAG/CUG repeat sequences without upstream AUG initiation codons using cellular models and patients' tissues of SCA8 and DM1, and named this RAN translation (Zu et al. 2011). Since this discovery, RAN translation has been pathologically confirmed to occur in patients' tissues of other noncoding repeat expansion diseases, including C9-ALS/FTD (Ash et al. 2013, Gendron et al. 2013, Mori et al. 2013a, Mori et al. 2013b, Zu et al. 2013), FXTAS (Todd et al. 2013), fragile X-associated primary ovarian insufficiency (Buijsen et al. 2016), DM2 (Zu et al. 2017), SCA31 (Ishiguro et al. 2017), SCA36 (McEachin et al. 2020, Todd et al. 2020), and Fuchs endothelial corneal dystrophy type 3 (Soragni et al. 2018) (Table 1.3). RAN translation is also observed in Huntington disease, one of the coding repeat expansion diseases (Bañez-Coronel et al. 2015).

RAN translation is characterized by unique features, including AUG codon-independent and repeat sequence-dependent translation initiation, translation from all reading frames, and translation upregulation by cellular stress (Figure 1.1), each of which is explained in detail as follows. First, RAN translation is initiated in an AUG codon-independent manner. Although the precise processes of RAN translation initiation remain to be fully elucidated, two distinct mechanisms have been proposed. Several studies have suggested that RAN translation is initiated from

Table 1.3 RAN translation in repeat expansion diseases

Disease	Gene	Location	Repeat motif	Putative RAN translation products	
				Sense	**Antisense**
C9orf72-linked amyotrophic lateral sclerosis/frontotemporal dementia	*C9orf72*	intron	GGGGCC	Gly-Ala*, Gly-Arg*, Gly-Pro*	Gly-Pro*, Pro-Ala*, Pro-Arg*
Fragile X-associated tremor/ataxia syndrome	*FMR1*	5′-UTR	CGG	Gly*, Ala, Arg	Ala*, Pro*, Arg
Fragile X-associated primary ovarian insufficiency	*FMR1*	5′-UTR	CGG	Gly*, Ala, Arg	Ala, Pro, Arg
Fuchs endothelial corneal dystrophy 3	*TCF4*	intron	CTG	Cys*, Ala, Leu	Gln*, Ala, Ser
Huntington disease	*HTT*	CDS	CAG	Ala*, Ser*, Gln	Cyc*, Leu*, Ala
Myotonic dystrophy 1	*DMPK*	3′-UTR	CTG	Ala, Cys, Leu	Gln*, Ala, Ser
Myotonic dystrophy 2	*ZNF9*	intron	CCTG	Leu-Pro-Ala-Cys*	Gln-Ala-Gly-Arg*
Spinocerebellar ataxia 8**	*ATXN8/ ATXN8OS*	CDS/ 3′-UTR	CAG/ CTG	(CAG) Ala*, Ser*, Gln	(CTG) Ala, Cys, Leu
Spinocerebellar ataxia 31	*BEAN*	intron	TGGAA	Trp-Asn-Gly-Met-Arg*, #	Ser-Ile-Pro-Phe-His
Spinocerebellar ataxia 36	*NOP56*	intron	GGCCTG	Gly-Pro*, Gly-Leu, Trp-Ala	Pro-Arg*, Gly-Pro*, Ala-Gln

*Products confirmed in patients.
**Caused by the expansion of repeat sequences within two overlapping genes (*ATXN8* and *ATXN8OS*), which are oppositely transcribed.
#Since the AUG sequence is included in the expanded repeat sequence itself, it is undetermined whether this product is generated by RAN translation or canonical AUG translation.
Abbreviations: CDS, coding sequence; UTR, untranslated region

the near-cognate initiation codons located upstream of the repeat sequence, requiring 5′ capping of the template RNA and ribosomal scanning, akin to physiological upstream open reading frame translation (Kearse et al. 2016, Green et al. 2017, Sellier et al. 2017, Tabet et al. 2018). For example, the CUG near-cognate codon upstream of the GGGGCC repeat sequences is utilized for RAN translation initiation in C9-ALS/FTD, as evidenced by *in vitro* translation assays and ribosome-profiling using patient-derived induced pluripotent stem cells (van 't Spijker et al. 2022). In contrast, other studies have indicated that RAN translation can be initiated directly within the repeat sequences in a 5′ cap-independent manner (Cheng et al. 2018), resembling noncanonical Internal Ribosomal Entry Site (IRES) translation, in which specific translation factors and ribosomal proteins are recruited directly onto the structured template RNA (Kwan and Thompson 2019). Second, RAN translation occurs in a repeat sequence-dependent manner. Although the thresholds of the repeat number required for translation initiation remain unknown, longer repeat sequences have been reported to show higher translation activity (Zu et al. 2011, Zu et al. 2013, Kearse et al. 2016, McEachin et al. 2020). In the first discovery of RAN translation in SCA8, the translation was shown to be initiated from the hairpin-forming CAG repeat RNA, but not from the non-hairpin-forming CAA repeat RNA (Zu et al. 2011), implying that higher-order structures formed by repeat RNA play important roles in RAN translation initiation. In C9-ALS/FTD, the G-quadruplex structure formed by the GGGGCC repeat RNA is considered to facilitate RAN translation initiation (Tabet et al. 2018). Third, RAN translation occurs in all reading frames, producing different repeat polypeptides from the same repeat sequences. In C9-ALS/FTD, for example, from the sense GGGGCC and antisense CCCCGG repeat RNA, RAN translation generates five distinct Dipeptide Repeat Proteins (DPRs); i.e., polyglycine-alanine (polyGA), polyglycine-arginine (polyGR), polyglycine-proline (polyGP), polyproline-alanine (polyPA) and polyproline-arginine (polyPR) (Ash et al. 2013, Gendron et al. 2013, Mori et al. 2013a, Mori et al. 2013b, Zu et al. 2013). Frameshifting during translation elongation can result in the production of chimeric repeat polypeptides, such as polyGA-GP in C9-ALS/FTD, which show different solubility and toxicity profiles compared with homomeric repeat polypeptides, thereby further contributing to the diversity of RAN-translated repeat polypeptides (McEachin et al. 2020, Wright et al. 2022, Latallo et al. 2023). Finally, RAN translation is upregulated by cellular stress, whereas global canonical translation is generally suppressed under stressed conditions through the integrated stress response, a stress-induced compensatory response characterized by the activation of several protein kinases and the subsequent phosphorylation of eukaryotic initiation factor (eIF) 2α (Green et al. 2017, Cheng et al. 2018, Sonobe et al. 2018, Westergard et al. 2019, Tusi et al. 2021). Importantly, RAN-translated repeat polypeptides themselves can induce oxidative and endoplasmic reticulum stress, and thus they may further upregulate RAN translation, potentially creating a feedforward loop that exacerbates the pathogenesis (Cheng et al. 2018).

The regulatory mechanisms governing RAN translation have gradually been uncovered, although much remains to be clarified. Several RNA helicases and RNA chaperones have been found to regulate RAN translation activity through modifying repeat RNA structures, such as the G-quadruplex structure (Ishiguro et al. 2017, Cheng et al. 2019, Liu et al. 2021, Tseng et al. 2021, Fujino et al. 2023b). In addition, specific translation factors, including eIFs and ribosomal proteins, have also been implicated in the regulation of RAN translation. For example, eIF3F and RPS25, translation factors facilitating IRES translation rather than canonical AUG-dependent translation, have been found to also upregulate RAN translation, suggesting a potential overlap in their regulatory mechanisms (Ayhan et al. 2018, Yamada et al. 2019).

ABERRANT LLPS CONDITIONS IN REPEAT EXPANSION DISEASES

The discovery of RAN translation has highlighted the common pathomechanisms driven by toxic repeat polypeptides in both coding and noncoding repeat expansion diseases. Recently, substantial attention has focused on the physiological LLPS of disease-linked proteins as a precursor of their pathological aggregation, and the perturbation of physiological LLPS by these repeat polypeptides.

LLPS is a thermodynamic phenomenon in which a solution separates into distinct phases rather than mixing homogeneously (Gomes and Shorter 2019). In biological contexts, LLPS contributes to the formation of dynamic biomolecular condensates known as MLOs in both the nucleus (e.g., nucleoli, Cajal bodies, nuclear speckles, and paraspeckles) and the cytoplasm (e.g., stress granules and P-bodies). MLOs compartmentalize specific proteins and RNAs to regulate a wide range of biological processes, including RNA metabolism, transcription regulation, DNA repair, and stress responses. Proteins incorporated into MLOs via LLPS often have a Low-Complexity Domain (LCD), a domain consisting of a few limited types of amino acids (Gomes and Shorter 2019). LCDs in proteins are known to act as Intrinsically Disordered Regions (IDRs) that do not adopt defined tertiary structures, thus facilitating multivalent interactions among proteins and RNAs through LLPS within MLOs

(Darling and Uversky 2017). Since amino acid repeat sequences also function as LCDs, they can take on dynamic conformations as IDRs, contributing to multivalent biomolecular interactions essential for physiological LLPS and MLOs formation (Darling and Uversky 2017). However, pathogenic expanded repeat polypeptides can aberrantly self-interact by themselves and also interact with other LCD-containing proteins, disrupting the dynamics of physiological LLPS, which in turn leads to their aggregation and the functional perturbation of MLOs.

In polyglutamine diseases, the *ATXN1* and *ATXN2* genes are the causative genes for SCA1 and SCA2 respectively, and wild-type ataxin-1 and ataxin-2 proteins are involved in the formation of MLOs, such as nucleoli and stress granules (Nonhoff et al. 2007, Zhang et al. 2020). Liquid droplets composed of polyglutamine proteins formed via LLPS can undergo phase transition into fibrous aggregates in a polyglutamine length-dependent manner both *in vitro* and in yeast (Peskett et al. 2018). Thus, wild-type proteins with normal-length polyglutamine tracts themselves are thought to participate physiologically in the formation of MLOs, whereas mutant proteins with expanded polyglutamine tracts induce aberrant phase transition and form aggregates, potentially disrupting MLO homeostasis.

In congenital polyalanine diseases, as discussed above, the causative genes encode transcription factors involved in development, which form transcriptional condensates, another type of nuclear MLO. The mutant proteins containing the expanded polyalanine tracts disrupt the dynamics and composition of these transcriptional condensates (Basu et al. 2020). Indeed, expanded polyalanine tracts in the mutant PHOX2B protein, which is associated with congenital central hypoventilation syndrome, can trigger a liquid-to-solid phase transition and sequester wild-type proteins into aberrant solid condensates, leading to the loss-of-function of PHOX2B (Antón et al. 2024).

RAN-translated repeat polypeptides also form liquid droplets via LLPS, as evidenced by the polyglycine protein in FXTAS (Asamitsu et al. 2021), and polyGR and polyPR proteins in C9-ALS/FTD (Boeynaems et al. 2017). These arginine-containing DPRs in C9-ALS/FTD can interact with other LCD-containing proteins, many of which are components of physiological MLOs, such as nucleolin and NPM1 in nucleoli and TIA-1 and G3BP1 in stress granules (Lee et al. 2016, Boeynaems et al. 2017). Consequently, DPRs may disrupt the dynamics and function of these MLOs.

In addition to repeat polypeptides, repeat RNA can also perturb physiological LLPS. Expanded repeat sequences in RNA strengthen multivalent RNA-RNA and RNA-RBP interactions. Indeed, CAG and CUG repeat RNA with more than 30 repeats form gel-like condensates *in vitro*. Moreover, short GGGGCC repeat RNA form spherical condensates, whereas long GGGGCC repeats form mesh-like aggregates, possibly owing to phase transition (Jain and Vale 2017). The higher-order structures formed by repeat RNA can also modify LLPS dynamics. For instance, the G-quadruplex structure of GGGGCC repeat RNA is reported to be important for condensate formation (Raguseo et al. 2023). In addition, the G-quadruplex-structured CGG repeat RNA can induce a liquid-to-solid phase transition of the polyglycine protein in FXTAS models (Asamitsu et al. 2021).

Of note, disease-linked proteins in sporadic neurodegenerative diseases, such as TDP-43, FUS, α-synuclein, and tau, also have LCDs and the potential to form condensates via LLPS *in vitro*. In cellular contexts, TDP-43 and FUS are essential components of stress granules and paraspeckles, respectively, and α-synuclein is suggested to be involved in P-bodies (Zbinden et al. 2020, Hallacli et al. 2022). Therefore, the perturbation of LLPS conditions of these proteins is also considered to be a key mechanism of their pathological aggregation that commonly underlies neurodegeneration. Intriguingly, pathological aggregation of these proteins is also observed in some repeat expansion diseases, such as TDP-43 accumulation in C9-ALS/FTD (DeJesus-Hernandez et al. 2011), *ATXN2*-lnked ALS (Elden et al. 2010) and *LRP12*-linked ALS (Kume et al. 2023), and α-synuclein accumulation in *ATXN2*-linked Parkinson's disease (Takao et al. 2011), implying potential molecular links between repeat polypeptides/RNA and these proteins through the dysregulation of LLPS, although the detailed mechanisms remain to be elucidated.

CONCLUSIONS

Recent advances in genome sequencing and analysis technologies have facilitated the discovery of novel repeat expansion diseases, and unveiled the involvement of expanded STRs even in sporadic diseases, such as ALS (Henden et al 2023), autism (Trost et al. 2020, Mitra et al. 2021), and cardiomyopathy (Mitina et al. 2024). Repeat polypeptides may contribute to the onset and progression of these diseases, serving as common pathomechanisms involved not only in the inherited repeat expansion diseases, but also in sporadic diseases. Therefore, understanding how repeat polypeptides are produced and exert toxic effects—particularly concerning the regulatory mechanisms of RAN translation and the aberrant LLPS conditions induced by repeat polypeptides—is crucial towards the therapeutic development across a broad range of these diseases.

ACKNOWLEDGEMENTS

We thank the members of our laboratory for helpful discussions, and Dr. H. Akiko Popiel for English editing of the manuscript.

ABBREVIATIONS

C9-ALS/FTD, *C9orf72*-linked amyotrophic lateral sclerosis/ frontotemporal dementia; DM, myotonic dystrophy; DPR, dipeptide repeat protein; eIF, eukaryotic initiation factor; FXTAS, fragile X tremor/ ataxia syndrome; FXS, fragile X syndrome; IDR, intrinsically disordered region; IRES, internal ribosomal entry site; LLPS, liquid-liquid phase separation; MLO, membraneless organelle; NIID, neuronal intranuclear inclusion disease; OPDM, oculopharyngodistal myopathy; polyGA, polyglycine-alanine; polyGP, polyglycine-proline; polyGR, polyglycine-arginine; polyPA, polyproline-alanine; polyPR, polyproline-arginine; RAN, repeat-associated non-AUG; RBP, RNA-binding protein; SBMA, spinal and bulbar muscular atrophy; SCA, spinocerebellar ataxia; STR, short tandem repeat.

REFERENCES

Albrecht, A. and S. Mundlos. 2005. The other trinucleotide repeat: polyalanine expansion disorders. Curr. Opin. Genet. Dev. 15: 285–293.

Antón, R., M.Á. Treviño, D. Pantoja-Uceda, S. Félix, M. Babu, E.J. Cabrita, et al. 2024. Alternative low-populated conformations prompt phase transitions in polyalanine repeat expansions. Nat. Commun. 15: 1925.

Asamitsu, S., Y. Yabuki, S. Ikenoshita, K. Kawakubo, M. Kawasaki, S. Usuki, et al. 2021. CGG repeat RNA G-quadruplexes interact with FMRpolyG to cause neuronal dysfunction in fragile X-related tremor/ataxia syndrome. Sci. Adv. 7: eabd9440.

Ash, P.E., K.F. Bieniek, T.F. Gendron, T. Caulfield, W.L. Lin, M. Dejesus-Hernandez, et al. 2013. Unconventional translation of *C9ORF72* GGGGCC expansion generates insoluble polypeptides specific to c9FTD/ ALS. Neuron. 77: 639–646.

Ayhan, F., B.A. Perez, H.K. Shorrock, T. Zu, M. Banez-Coronel, T. Reid, et al. 2018. SCA8 RAN polySer protein preferentially accumulates in white matter regions and is regulated by eIF3F. EMBO J. 37: e99023.

Bañez-Coronel, M., F. Ayhan, A.D. Tarabochia, T. Zu, B.A. Perez, S.K. Tusi, et al. 2015. RAN translation in Huntington disease. Neuron. 88: 667–677.

Basu, S., S.D. Mackowiak, H. Niskanen, D. Knezevic, V. Asimi, S. Grosswendt, et al. 2020. Unblending of transcriptional condensates in human repeat expansion disease. Cell. 181: 1062–1079.

Boeynaems, S., E. Bogaert, D. Kovacs, A. Konijnenberg, E. Timmerman, A. Volkov, et al. 2017. Phase separation of *C9orf72* dipeptide repeats perturbs stress granule dynamics. Mol. Cell. 65: 1044–1055.

Boivin, M., J. Deng, V. Pfister, E. Grandgirard, M. Oulad-Abdelghani, B. Morlet, et al. 2021. Translation of GGC repeat expansions into a toxic polyglycine protein in NIID defines a novel class of human genetic disorders: The polyG diseases. Neuron. 109: 1825–1835.

Buijsen, R.A., J.A. Visser, P. Kramer, E.A. Severijnen, M. Gearing, N. Charlet-Berguerand, et al. 2016. Presence of inclusions positive for polyglycine containing protein, FMRpolyG, indicates that repeat-associated non-AUG translation plays a role in fragile X-associated primary ovarian insufficiency. Hum. Reprod. 31: 158–168.

Bürk, K. 2017. Friedreich Ataxia: current status and future prospects. Cerebellum Ataxias. 4:4.

Chen, Y.C., M. Auer-Grumbach, S. Matsukawa, M. Zitzelsberger, A.C. Themistocleous, T.M. Strom, et al. 2015. Transcriptional regulator PRDM12 is essential for human pain perception. Nat. Genet. 47: 803–808.

Chen, Z., E.K. Gustavsson, H. Macpherson, C. Anderson, C. Clarkson, C. Rocca, et al. 2024. Adaptive long-read sequencing reveals GGC repeat expansion in *ZFHX3* associated with spinocerebellar ataxia type 4. Mov. Disord. 39: 486–497.

Cheng, W., S. Wang, A.A. Mestre, C. Fu, A. Makarem, F. Xian, et al. 2018. C9ORF72 GGGGCC repeat-associated non-AUG translation is upregulated by stress through eIF2α phosphorylation. Nat. Commun. 9: 51.

Cheng, W., S. Wang, Z. Zhang, D.W. Morgens, L.R. Hayes, S. Lee, et al. 2019. CRISPR-Cas9 screens identify the RNA helicase DDX3X as a repressor of *C9ORF72* (GGGGCC)n repeat-associated non-AUG translation. Neuron. 104: 885–898.

Dalski, A., M.G. Pauly, H. Hanssen, J. Hagenah, Y. Hellenbroich, C. Schmidt, et al. 2024. Repeat length in spinocerebellar ataxia type 4 (SCA4) predicts age at onset and disease severity. J. Neurol. 271: 6289–6300.

Darling, A.L. and V.N. Uversky. 2017. Intrinsic disorder in proteins with pathogenic repeat expansions. Molecules. 22: 2027.

DeJesus-Hernandez, M., I.R. Mackenzie, B.F. Boeve, A.L. Boxer, M. Baker, N.J. Rutherford, et al. 2011. Expanded GGGGCC hexanucleotide repeat in noncoding region of *C9ORF72* causes chromosome 9p-linked FTD and ALS. Neuron. 72: 245–256.

Elden, A.C., H.J. Kim, M.P. Hart, A.S. Chen-Plotkin, B.S. Johnson, X. Fang, et al. 2010. Ataxin-2 intermediate-length polyglutamine expansions are associated with increased risk for ALS. Nature. 466: 1069–1075.

Figueroa, K.P., C. Gross, E. Buena-Atienza, S. Paul, M. Gandelman, N. Kakar, et al. 2024. A GGC-repeat expansion in *ZFHX3* encoding polyglycine causes spinocerebellar ataxia type 4 and impairs autophagy. Nat. Genet. 56: 1080–1089.

Fujino, Y. and Y. Nagai. 2022. The molecular pathogenesis of repeat expansion diseases. Biochem. Soc. Trans. 50: 119–134.

Fujino, Y., K. Mori and Y. Nagai. 2023a. Repeat-associated non-AUG translation in neuromuscular diseases: mechanisms and therapeutic insights. J. Biochem. 173: 273–281.

Fujino, Y., M. Ueyama, T. Ishiguro, D. Ozawa, H. Ito, T. Sugiki, et al. 2023b. FUS regulates RAN translation through modulating the G-quadruplex structure of GGGGCC repeat RNA in *C9orf72*-linked ALS/FTD. eLife. 12: RP84338.

Gendron, T.F., K.F. Bieniek, Y.J. Zhang, K. Jansen-West, P.E. Ash, T. Caulfield, et al. 2013. Antisense transcripts of the expanded *C9ORF72* hexanucleotide repeat form nuclear RNA foci and undergo repeat-associated non-ATG translation in c9FTD/ALS. Acta. Neuropathol. 126: 829–844.

Gomes, E. and J. Shorter. 2019. The molecular language of membraneless organelles. J. Biol. Chem. 294: 7115–7127.

Goodman, F.R., S. Mundlos, Y. Muragaki, D. Donnai, M.L. Giovannucci-Uzielli, E. Lapi, et al. 1997. Synpolydactyly phenotypes correlate with size of expansions in HOXD13 polyalanine tract. Proc. Natl. Acad. Sci. USA. 94: 7458–7463.

Green K.M., M.R. Glineburg, M.G. Kearse, B.N. Flores, A.E. Linsalata, S.J. Fedak, et al. 2017. RAN translation at *C9orf72*-associated repeat expansions is selectively enhanced by the integrated stress response. Nat. Commun. 8: 2005.

Hallacli, E., C. Kayatekin, S. Nazeen, X.H. Wang, Z. Sheinkopf, S. Sathyakumar, et al. 2022. The Parkinson's disease protein alpha-synuclein is a modulator of processing bodies and mRNA stability. Cell. 185: 2035–2056.

Henden, L., L.G. Fearnley, N. Grima, E.P. McCann, C. Dobson-Stone, L. Fitzpatrick, et al. 2023. Short tandem repeat expansions in sporadic amyotrophic lateral sclerosis and frontotemporal dementia. Sci. Adv. 9: eade2044.

Ishiguro, T., N. Sato, M. Ueyama, N. Fujikake, C. Sellier, A. Kanegami, et al. 2017. Regulatory role of RNA chaperone TDP-43 for RNA misfolding and repeat-associated translation in SCA31. Neuron. 94: 108–124.

Ishihara, T., M. Tada, Y. Kanemitsu, Y. Takahashi, K. Ishikawa, K. Ikenaka, et al. 2024. L-arginine in patients with spinocerebellar ataxia type 6: a multicentre, randomised, double-blind, placebo-controlled, phase 2 trial. EClinicalMedicine. 78: 102952.

Jain, A. and R. Vale. 2017. RNA phase transitions in repeat expansion disorders. Nature. 546: 243–247.

Kato, M., S. Das, K. Petras, K. Kitamura, K.I. Morohashi, D.N. Abuelo, et al. 2004. Mutations of *ARX* are associated with striking pleiotropy and consistent genotype-phenotype correlation. Hum. Mutat. 23: 147–159.

Kearse, M.G., K.M. Green, A. Krans, C.M. Rodriguez, A.E. Linsalata, A.C. Goldstrohm, et al. 2016. CGG repeat-associated non-AUG translation utilizes a cap-dependent scanning mechanism of initiation to produce toxic proteins. Mol. Cell. 62: 314–322.

Kremer, E.J., M. Pritchard, M. Lynch, S. Yu, K. Holman, E. Baker, et al. 1991. Mapping of DNA instability at the fragile X to a trinucleotide repeat sequence p(CCG)n. Science. 252: 1711–1714.

Kume, K., T. Kurashige, K. Muguruma, H. Morino, Y. Tada, M. Kikumoto, et al. 2023. CGG repeat expansion in *LRP12* in amyotrophic lateral sclerosis. Am. J. Hum. Genet. 110: 1086–1097.

Kwan, T. and S.R. Thompson. 2019. Noncanonical translation initiation in eukaryotes. Cold Spring Harb. Perspect. Biol. 11: a032672.

La Spada, A.R., E.M. Wilson, D.B. Lubahn, A.E. Harding and K.H. Fischbeck. 1991. Androgen receptor gene mutations in X-linked spinal and bulbar muscular atrophy. Nature. 352: 77–79.

Latallo, M.J., S. Wang, D. Dong, B. Nelson, N.M. Livingston, R. Wu, et al. 2023. Single-molecule imaging reveals distinct elongation and frameshifting dynamics between frames of expanded RNA repeats in C9ORF72-ALS/FTD. Nat. Commun. 14: 5581.

Lee, K.H., P. Zhang, H.J. Kim, D.M. Mitrea, M. Sarkar, B.D. Freibaum, et al. 2016. C9orf72 dipeptide repeats impair the assembly, dynamics, and function of membrane-less organelles. Cell. 167: 774–788.

Lin, X., J.W. Miller, A. Mankodi, R.N. Kanadia, Y. Yuan, R.T. Moxley, et al. 2006. Failure of MBNL1-dependent post-natal splicing transitions in myotonic dystrophy. Hum. Mol. Genet. 15: 2087–2097.

Liu, H., Y.N. Lu, T. Paul, G. Periz, M.T. Banco, A.R. Ferré-D'Amaré, et al. 2021. A helicase unwinds hexanucleotide repeat RNA G-quadruplexes and facilitates repeat-associated non-AUG translation. J. Am. Chem. Soc. 143: 7368–7379.

Malik, I., C.P. Kelley, E.T. Wang and P.K. Todd. 2021. Molecular mechanisms underlying nucleotide repeat expansion disorders. Nat. Rev. Mol. Cell. Biol. 22: 589–607.

Mankodi, A., M.P. Takahashi, H. Jiang, C.L. Beck, W.J. Bowers, R.T. Moxley, et al. 2002. Expanded CUG repeats trigger aberrant splicing of ClC-1 chloride channel pre-mRNA and hyperexcitability of skeletal muscle in myotonic dystrophy. Mol. Cell. 10: 35–44.

Matera, I., T. Bachetti, F. Puppo, M. Di Duca, F. Morandi, G.M. Casiraghi, et al. 2004. *PHOX2B* mutations and polyalanine expansions correlate with the severity of the respiratory phenotype and associated symptoms in both congenital and late onset Central Hypoventilation syndrome. J. Med. Genet. 41: 373–380.

McEachin, Z.T., T.F. Gendron, N. Raj, M. García-Murias, A. Banerjee, R.H. Purcell, et al. 2020. Chimeric peptide species contribute to divergent dipeptide repeat pathology in c9ALS/FTD and SCA36. Neuron. 107: 292–305.

Minakawa, E.N., H.A. Popiel, M. Tada, T. Takahashi, H. Yamane, Y. Saitoh, et al. 2020. Arginine is a disease modifier for polyQ disease models that stabilizes polyQ protein conformation. Brain. 143: 1811–1825.

Mitina, A., M. Khan, R. Lesurf, Y. Yin, W. Engchuan, O. Hamdan, et al. 2024. Genome-wide enhancer-associated tandem repeats are expanded in cardiomyopathy. EBioMedicine. 101: 105027.

Mitra, I., B. Huang, N. Mousavi, N. Ma, M. Lamkin, R. Yanicky, et al. 2021. Patterns of de novo tandem repeat mutations and their role in autism. Nature. 589: 246–250.

Mori, K., S.M. Weng, T. Arzberger, S. May, K. Rentzsch, E. Kremmer, et al. 2013a. The *C9orf72* GGGGCC repeat is translated into aggregating dipeptide-repeat proteins in FTLD/ALS. Science. 339: 1335–1338.

Mori, K., T. Arzberger, F.A. Grässer, I. Gijselinck, S. May, K. Rentzsch, et al. 2013b. Bidirectional transcripts of the expanded *C9orf72* hexanucleotide repeat are translated into aggregating dipeptide repeat proteins. Acta Neuropathol. 126: 881–893.

Myrick, L.K., M. Nakamoto-Kinoshita, N.M. Lindor, S. Kirmani, X. Cheng and S.T. Warren 2014. Fragile X syndrome due to a missense mutation. Eur. J. Hum. Genet. 22: 1185–1189.

Nagai, Y., T. Inui, H.A. Popiel, N. Fujikake, K. Hasegawa, Y. Urade, et al. 2007. A toxic monomeric conformer of the polyglutamine protein. Nat. Struct. Mol. Biol. 14: 332–340.

Nonhoff, U., M. Ralser, F. Welzel, I. Piccini, D. Balzereit, M.L. Yaspo, et al. 2007. Ataxin-2 interacts with the DEAD/H-box RNA helicase DDX6 and interferes with P-bodies and stress granules. Mol. Biol. Cell. 18: 1385–1396.

Paucar, M., D. Nilsson, M. Engvall, J. Laffita-Mesa, C. Söderhäll, M. Skorpil, et al. 2024. Spinocerebellar ataxia type 4 is caused by a GGC expansion in the *ZFHX3* gene and is associated with prominent dysautonomia and motor neuron signs. J. Intern. Med. 296: 234–248.

Pérez Baca, M.D.R., E.Z. Jacobs, L. Vantomme, P. Leblanc, E. Bogaert, A. Dheedene, et al. 2024. Haploinsufficiency of *ZFHX3*, encoding a key player in neuronal development, causes syndromic intellectual disability. Am. J. Hum. Genet. 111: 509–528.

Peskett, T.R., F. Rau, J. O'Driscoll, R. Patani, A.R. Lowe and H.R. Saibil 2018. A liquid to solid phase transition underlying pathological huntingtin exon1 aggregation. Mol. Cell. 70: 588–601.

Raguseo, F., Y. Wang, J. Li, M. Petrić Howe, R. Balendra, A. Huyghebaert, et al. 2023. The ALS/FTD-related *C9orf72* hexanucleotide repeat expansion forms RNA condensates through multimolecular G-quadruplexes. Nat. Commun. 14: 8272.

Sellier, C., R. Buijsen, F. He, S. Natla, L. Jung, P. Tropel, et al. 2017. Translation of expanded CGG repeats into FMRpolyG is pathogenic and may contribute to Fragile X tremor ataxia syndrome. Neuron. 93: 331–347.

Sonobe, Y., G. Ghadge, K. Masaki, A. Sendoel, E. Fuchs and R.P. Roos. 2018. Translation of dipeptide repeat proteins from the C9ORF72 expanded repeat is associated with cellular stress. Neurobiol. Dis. 116: 155–165.

Soragni, E., L. Petrosyan, T.A. Rinkoski, E.D. Wieben, K.H. Baratz, M.P. Fautsch, et al. 2018. Repeat-associated non-ATG (RAN) translation in Fuchs' endothelial corneal dystrophy. Invest. Ophthalmol. Vis. Sci. 59: 1888–1896.

Tabet, R., L. Schaeffer, F. Freyermuth, M. Jambeau, M. Workman, C.Z. Lee, et al. 2018. CUG initiation and frameshifting enable production of dipeptide repeat proteins from ALS/FTD C9ORF72 transcripts. Nat. Commun. 9: 152.

Takao, M., M. Aoyama, K. Ishikawa, Y. Sakiyama, H. Yomono, Y. Saito, et al. 2011. Spinocerebellar ataxia type 2 is associated with Parkinsonism and Lewy body pathology. BMJ Case Rep. bcr0120113685.

Takeuchi, T. and Y. Nagai. 2017. Protein misfolding and aggregation as a therapeutic target for polyglutamine diseases. Brain. Sci. 7: 128.

Tan, D., C. Wei, Z. Chen, Y. Huang, J. Deng, J. Li, et al. 2023. CAG repeat expansion in *THAP11* is associated with a novel spinocerebellar ataxia. Mov. Disord. 38: 1282–1293.

Todd, P.K., S.Y. Oh, A. Krans, F. He, C. Sellier, M. Frazer, et al. 2013. CGG repeat-associated translation mediates neurodegeneration in fragile X tremor ataxia syndrome. Neuron. 78: 440–455.

Todd, T.W., Z.T. McEachin, J. Chew, A.R. Burch, K. Jansen-West, J. Tong, et al. 2020. Hexanucleotide repeat expansions in c9FTD/ALS and SCA36 confer selective patterns of neurodegeneration *in vivo*. Cell. Rep. 31: 107616.

Trost, B., W. Engchuan, C.M. Nguyen, B. Thiruvahindrapuram, E. Dolzhenko, I. Backstrom, et al. 2020. Genome-wide detection of tandem DNA repeats that are expanded in autism. Nature. 586: 80–86.

Tseng, Y.J., S.N. Sandwith, K.M. Green, A.E. Chambers, A. Krans, H.M. Raimer, et al. 2021. The RNA helicase DHX36/G4R1 modulates C9orf72 GGGGCC hexanucleotide repeat- associated translation. J. Biol. Chem. 297: 100914.

Tusi, S.K., L. Nguyen, K. Thangaraju, J. Li, J.D. Cleary, T. Zu, et al. 2021. The alternative initiation factor eIF2A plays key role in RAN translation of myotonic dystrophy type 2 CCUG•CAGG repeats. Hum. Mol. Genet. 30: 1020–1029.

van 't Spijker, H.M., E.E. Stackpole, S. Almeida, O. Katsara, B. Liu, K. Shen, et al. 2022. Ribosome profiling reveals novel regulation of *C9ORF72* GGGGCC repeat-containing RNA translation. RNA. 28: 123–138.

Wallenius, J., E. Kafantari, E. Jhaveri, S. Gorcenco, A. Ameur, C. Karremo, et al. 2024. Exonic trinucleotide repeat expansions in *ZFHX3* cause spinocerebellar ataxia type 4: A poly-glycine disease. Am. J. Hum. Genet. 111: 82–95.

Westergard, T., K. McAvoy, K. Russell, X. Wen, Y. Pang, B. Morris, et al. 2019. Repeat-associated non-AUG translation in C9orf72-ALS/FTD is driven by neuronal excitation and stress. EMBO Mol. Med. 11: e9423.

Wright, S.E., C.M. Rodriguez, J. Monroe, J. Xing, A. Krans, B.N. Flores, et al. 2022. CGG repeats trigger translational frameshifts that generate aggregation-prone chimeric proteins. Nucleic Acids Res. 50: 8674–8689.

Yamada, S.B., T.F. Gendron, T. Niccoli, N.R. Genuth, R. Grosely, Y. Shi, et al. 2019. RPS25 is required for efficient RAN translation of C9orf72 and other neurodegenerative disease-associated nucleotide repeats. Nat. Neurosci. 22: 1383–1388.

Yu, J., J. Deng, X. Guo, J. Shan, X. Luan, L. Cao, et al. 2021. The GGC repeat expansion in *NOTCH2NLC* is associated with oculopharyngodistal myopathy type 3. Brain. 144: 1819–1832.

Yu, J., J. Shan, M. Yu, L. Di, Z. Xie, W. Zhang, et al. 2022. The CGG repeat expansion in *RILPL1* is associated with oculopharyngodistal myopathy type 4. Am. J. Hum. Genet. 109: 533–541.

Zbinden, A., M. Pérez-Berlanga, P. De Rossi and M. Polymenidou 2020. Phase separation and neurodegenerative diseases: a disturbance in the force. Dev. Cell. 55: 45–68.

Zhang, S., E. Hinde, M. Parkyn Schneider, D.A. Jans and M.A. Bogoyevitch. 2020. Nuclear bodies formed by polyQ-ataxin-1 protein are liquid RNA/protein droplets with tunable dynamics. Sci. Rep. 10: 1557.

Zhong, S., Y. Lian, W. Luo, R. Luo, X. Wu, J. Ji, et al. 2021. Upstream open reading frame with *NOTCH2NLC* GGC expansion generates polyglycine aggregates and disrupts nucleocytoplasmic transport: implications for polyglycine diseases. Acta Neuropathol. 142: 1003–1023.

Zu, T., B. Gibbens, N.S. Doty, M. Gomes-Pereira, A. Huguet, M.D. Stone, et al. 2011. Non-ATG-initiated translation directed by microsatellite expansions. Proc. Natl. Acad. Sci. USA. 108: 260–265.

Zu, T., Y. Liu, M. Bañez-Coronel, T. Reid, O. Pletnikova, J. Lewis, et al. 2013. RAN proteins and RNA foci from antisense transcripts in *C9ORF72* ALS and frontotemporal dementia. Proc. Natl. Acad. Sci. USA. 110: E4968–E4977.

Zu, T., J.D. Cleary, Y. Liu, M. Bañez-Coronel, J.L. Bubenik, F. Ayhan, et al. 2017. RAN translation regulated by muscleblind proteins in myotonic dystrophy type 2. Neuron. 95: 1292–1305.

Chapter 2

Pathologies Due to PolyQ Sequences

Raffaella Parlato and Patrick C.A. van der Wel*
Zernike Institute for Advanced Materials, University of Groningen, Nijenborgh 3, 9747 AG, Groningen, The Netherlands

INTRODUCTION

A category of fatal and incurable diseases is not caused by external agents like bacteria or viruses but is instead associated with the aggregation of endogenous proteins. In such cases, the affected proteins lose their structure and function, accumulating to form insoluble aggregates (Chiti and Dobson 2017). The family of diseases known as amyloid diseases includes Huntington's Disease (HD), Parkinson's Disease (PD), Alzheimer's disease, and spinocerebellar ataxias to name a few. Remarkably, these diseases involve different proteins, but they all tend to form a similar type of β-sheet-based aggregates known as amyloid or amyloid-like fibrils. Such amyloids have common features, such as the ability to bind special dyes like Congo red. Moreover, X-ray diffraction studies revealed a shared diffraction pattern termed cross-β (Iadanza et al. 2018). Today, the amyloid disease family encompasses 50 proteins associated with various disorders. These diseases can be categorized into three groups: neurodegenerative (affecting brain function like PD and HD), non-neuropathic localized (affecting specific tissues, e.g.,

*For Correspondence: p.c.a.van.der.wel@rug.nl

type II diabetes), and non-neuropathic systemic disorders (affecting the entire body like familial amyloid cardiomyopathy) (Chiti and Dobson 2017, Iadanza et al. 2018).

This chapter focuses on one remarkable class of protein aggregation diseases, in which proteins with an expanded repetitive polyglutamine (polyQ) segment undergo aggregation (Chiti and Dobson 2017). This phenomenon derives from the presence of uninterrupted polyQ segments in various proteins in the human body, reflecting genes with CAG codon repeats that can undergo pathogenic expansion. Multiple neurodegenerative disorders are characterized by expansions of these CAG repeats (or polyQ regions), affecting different genes (or proteins). When these expansions exceed a critical threshold, the affected individual becomes prone to some untreatable neurodegenerative diseases. Diverse mechanisms have been implicated in the expansion process, including DNA replication error, such as slippage of DNA polymerase, as well as multiple DNA repair mechanisms, both with possible connections to oxidative DNA damage (McMurray 2010, Liu and Wilson 2012). Although the expansion of polyQ repeats is linked to these lethal diseases, polyQ tracts are prevalent across the genomes of diverse species, suggesting a physiological role in their wild-type form (Schaefer et al. 2012). At the same time, homologous proteins in different species may entirely lack the polyQ tract yet retain their functional roles, implying that the polyQ tract is not indispensable. The polyQ tract has been implicated in mediating protein-protein interactions, such as those involving the transcriptional machinery (Schaefer et al. 2012).

Today, the polyQ-related syndromes encompass ten neurodegenerative diseases: Spinal-Bulbar Muscular Atrophy (SBMA), spinocerebellar ataxia (SCA) types 1, 2, 3, 6, 7, 17, Huntington Disease-Like 2 (HDL2), HD, and dentatorubral-pallidoluysian atrophy (DRPLA) (Johnson et al. 2022). Each disease stems from mutations in a specific protein. Remarkably, these proteins vary in their cellular roles, and in many cases, the pathways and interactions involving these proteins are not yet fully understood (Figures 2.1a and b). A common feature among these diseases is the inverse relationship between the CAG repeat (expansion) length and the age of onset: longer polyQ stretches lead to an earlier onset of disease (Figure 2.1c) (Wanker 2000, Kuiper et al. 2017). The length of the polyQ region varies among the wild-type proteins. However, for all of them, if the region becomes expanded beyond a certain threshold, it leads to disease manifestation. For most diseases, the threshold is around 36 repetitions, however, certain diseases have a lower threshold of 20 repetitions (Wanker 2000).

A seemingly common characteristic among the mutant proteins is their tendency to form aggregates, which correlates with the length of the (expanded) polyQ region. This aggregation propensity connects to

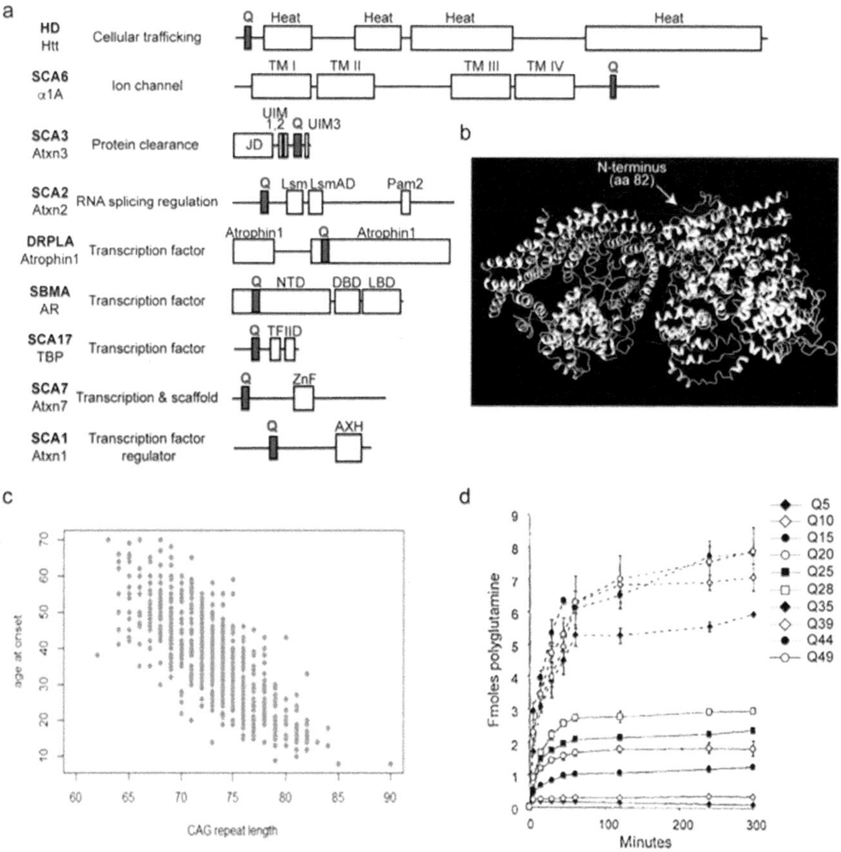

Figure 2.1 Polyglutamine-related disease overview. (a) Schematic representation of the nine proteins associated with the polyQ diseases: Spinal-Bulbar Muscular Atrophy (SBMA), spinocerebellar ataxia (SCA) types 1, 2, 3, 6, 7, 17, Huntington Disease (HD), and dentatorubral-pallidoluysian atrophy (DRPLA). Each polyQ disorder is denoted by the disease name in bold, the corresponding protein name, and the protein role on the left. The protein domain is represented with a box and the domain's name above. A darkened box represents the polyQ segment position, labeled with 'Q'. Adapted with permission from (Johnson et al. 2022). (b) In the center, the huntingtin (HTT) structure is depicted in white generated using PyMOL and PDB file 6YEJ. The arrow indicates the resolved N-terminus. Residues from HTTex1 are missing due to the high flexibility. (c) Graph showing the correlation between the polyQ length and the age of onset. Increasing the number of repetitions decreases the age of onset. Adapted with permission from (Akçimen et al. 2020). (d) Graph representing the correlation between the number of repetitions and increased aggregation propensity. Adapted with permission from (Chen et al. 2001).

the observation of protein aggregates or inclusions in the (post-mortem) brain tissue of patients suffering from diseases like HD. As discussed in more detail below, the role of these aggregated proteins in these diseases is under continued debate. Some studies indicate that such aggregates have pathogenic effects, while others argue that (certain types of) aggregates may serve as a protective mechanism to diminish toxic effects. In any case, aggregates are often seen as crucial hallmarks or biomarkers of disease and disease progression. As such, understanding their molecular structure and the aggregation process is crucial for elucidating disease mechanisms, designing novel treatment strategies, and developing aggregate-based biomarkers. To better understand the aggregation process, aggregation kinetics assays can be conducted. Such mechanistic studies allow an analysis of different features that influence the aggregation kinetics, such as the polyQ length and the flanking regions (Figure 2.1d, more below) (Chen et al. 2001).

HUNTINGTON'S DISEASE

After the above brief introduction about polyQ-aggregation-related syndromes more broadly, we narrow our focus to HD. It is the most common and also best understood among the polyQ or CAG-repeat expansion disorders. It was described for the first time by George Huntington in the 1870s. He studied several patient generations displaying identical symptoms, including involuntary muscle contractions causing dance-like movements. Initially, he reported the disease as confined to a few families (Stevenson 1934). It was later defined as a rare inherited disorder that causes progressive loss of function and death of neurons (brain cells) in specific brain regions, particularly the striatum and neocortex, areas responsible for movement and mood regulation. Damage in these regions leads to the onset of symptoms (Reiner et al. 2011, Waldvogel et al. 2015). As with other polyQ disorders, the first symptoms appear in middle age (~40 years old) or, in some cases, earlier (juvenile HD), depending on the length of the polyQ segment (Bates et al. 2015, Kuiper et al. 2017).

HD is an autosomal dominant neurodegenerative pathology, meaning it affects males and females equally and requires only one affected parent to transmit the disease. HD is caused by a mutation in the first exon (Ex1) of the huntingtin (HTT) gene situated in chromosome 4 (Bates et al. 2015). This chromosome was identified in 1983, while the gene was found in 1993 (Macdonald 1993). DNA cloning of the affected and healthy individuals revealed that the former had an expanded trinucleotide repeat in the HTT gene (Macdonald 1993). The mutation represents an expansion of an uninterrupted repeat of CAG codons, which codes for the polyQ segment in the HTT protein (Landles and Bates 2004, Bates et al. 2015). When the repeat length exceeds 35 CAG codons (or glutamines), the disease manifests—notably, the disease onset is earlier as the polyQ length increases (Figure 2.1b) (Wanker 2000, Chen et al. 2001). Despite identifying the mutated gene, the exact cause of the symptoms remained unclear.

In 1997, neuronal brain cells were imaged with anti-HTT antibodies, revealing intracellular inclusion (Figure 2.2a) (DiFiglia et al. 1997, Miguez et al. 2023) with a filamentous shape (Figure 2.2b) (Galaz-Montoya et al. 2021). Analogous inclusions or aggregates were subsequently also seen for other polyQ diseases (Jayaraman et al. 2009). Furthermore, using the antibody labeling, it was also clear that the HD aggregates were not made of the complete HTT protein but only by an N-terminal fragment, seemingly the first exon (HTTex1) of HTT (DiFiglia et al. 1997). At the time, it was found that the HTT with expanded polyQ tracts became more prone to proteolytic cleavage, producing (aggregation-prone) fragments. More recently, studies have highlighted parallel or competing mechanisms for generating N-terminal HTT fragments in patient cells and tissues through aberrant splicing of the mutant gene product (Neueder et al. 2017, Hoschek et al. 2024). These mechanisms are thought to produce HTTex1, which has been studied extensively in various contexts: *in vitro*, in cells, in animals, in patients, and in numerous simulation studies. Notably, recent work suggests that the propensity for HTTex1 generation correlates to the extent of CAG repeat expansion (Hoschek et al. 2024). Additionally, it is important to mention the evidence for somatic expansion in HD patients leading to the further expansion of CAG repeats over the patient's lifetime (Kennedy 2003, Mouro Pinto et al. 2020). The mechanism of somatic expansion is thought to involve DNA repair enzymes, seemingly explaining their discovery as genetic factors in HD. As the repeat length expands with age, the aberrant splicing would also increase, resulting in the generation of ever more aggregation-prone HTTex1 fragments bearing the expanding polyQ segment.

FEATURES AND BEHAVIOR OF THE POLYQ PROTEIN IN HD

HTT is a 350 kDa protein consisting of over 3,000 amino acids, and it is expressed throughout the body with particularly high levels in the brain (Schulte and Littleton 2011). In its wild-type form, HTT is soluble and highly conserved across different animal species through evolution (Saudou and Humbert 2016). Although its cellular function is not entirely elucidated, HTT appears to be involved in various neuronal

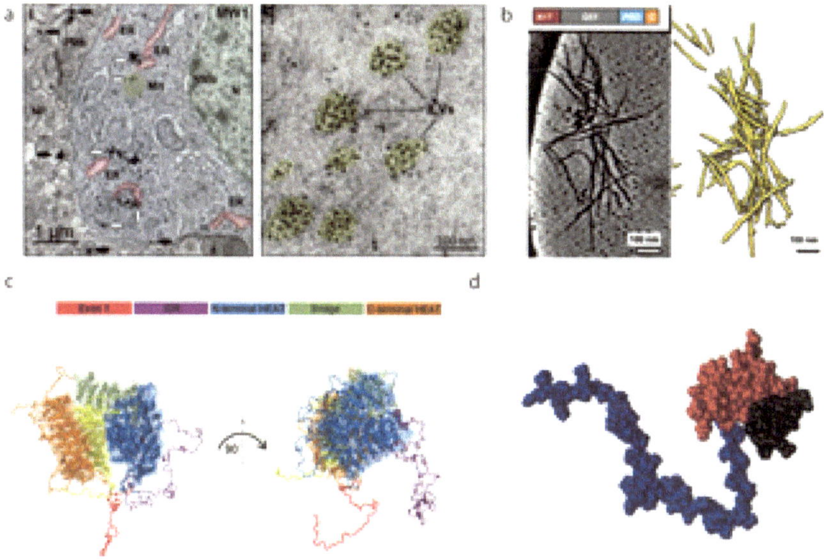

Figure 2.2 PolyQ aggregates images and monomeric structure of HTT protein. (a) Electron microscopy image of a brain cell with antibody labeling, illustrating extracellular (black arrows) and intracellular inclusion (white arrows). Adapted with permission from (Miguez et al. 2023). (b) Cryo-ET tomograph of mutated HTTex1 with a 51-residue polyQ segment. Adapted with permission from (Galaz-Montoya et al. 2021). (c) Surface representation of the EM structure of the HTT protein in soluble form. Adapted with permission from (Harding et al. 2019). (d) Tadpole-like structure of the HTTex1 fragment in its monomeric form. Adapted with permission from (Newcombe et al. 2018).

processes, such as cellular dynamics, metabolism, gene expression, and pathways critical for cell survival (Parsons and Raymond 2015, Saudou and Humbert 2016). Depletion of HTT is fatal for embryos, leading to critical brain defects, indicating it is essential in brain development (Schulte and Littleton 2011). Previous structural studies have shown that HTT protein contains multiple α-helical HEAT domains crucial for protein-protein interactions. However, several regions are Intrinsically Disordered Regions (IDRs), and hence, HTT is classified as an Intrinsically Disordered Protein (IDP). Consequently, these dynamic regions remain unresolved in available cryo-EM structures of full-length HTT (Saudou and Humbert 2016). Due to HTT's relative dynamism in isolation, structural analysis by cryo-EM was enabled by co-expression with HAP40, an HTT-binding protein (Figure 2.2c). Notably, the IDRs, such as the HTT N-terminal domain, are functionally important as they are the targets of Post-Translational Modification (PTMs) and significant for the role of HTT as a scaffold protein (Harding et al. 2019).

HTTex1 normally forms the N-terminal region of the large HTT protein, and it comprises three recognized segments: N17, the polyQ domain, and the Proline-Rich Domain (PRD) (DiFiglia et al. 1997). The first domain, N17 or HTT^{NT}, is located at the N-terminus and is 17 residues long (when one includes the initial methionine that is commonly processed *in vivo*). The N17 segment is essential for protein localization and membrane interaction (Atwal et al. 2007, Tao et al. 2019). Additionally, it harbors residues that undergo PTMs that can modulate aggregation and toxicity. The second domain is the polyQ segment itself. It varies in length and can include 10 to 35 glutamines in healthy individuals but is expanded beyond 35 in HD gene carriers. The third domain is the C-terminal region of exon 1, and it is highly proline-rich, hence the designation as PRD (Parsons and Raymond 2015). The PRD segment plays a role in HTT-protein interactions (Falk et al. 2020).

Generally speaking, HTTex1 is found to be highly dynamic, flexible, and disordered (Matlahov and Van der Wel 2019), qualifying it as an IDR and rendering it invisible in cryo-EM structures of full-length HTT. Indeed, it is believed that the natural role of (non-expanded) polyQ segments likely relates to their native disorder, coupled with a propensity to adopt more ordered structures. Intermolecular interactions can stabilize the formation of α-helical bundles, which may explain the tendency of AlphaFold models to propose helical conformations for polyQ segments. According to Förster Resonance Energy Transfer (FRET) experiments and atomistic simulations, the monomeric HTTex1 exhibits a tadpole-like structure in isolation (Newcombe et al. 2018). This structure consists of a rounded head comprising the helical N17 and the disordered but collapsed polyQ domain, and a more extended semiflexible PRD tail (Figure 2.2d) (Newcombe et al. 2018). This image illustrates the general observation that un-aggregated polyQ segments in solution may be highly dynamic, but they appear to be prone to interactions that lead to a somewhat compact disordered conformation.

AGGREGATION BY POLYQ SEGMENTS AND POLYQ PROTEINS

As mentioned above, the formation of aggregates or protein inclusions is a hallmark of neurodegenerative and polyQ disorders such as HD. Consequently, there is considerable interest in the mechanism of aggregation, as well as the structure and properties of the polyQ protein aggregates. In post-mortem patient tissues, polyQ proteins are observed in large inclusions within neuronal cells in brain regions affected by neurodegeneration (DiFiglia et al. 1997). The nature of these inclusions has been studied in more detail in cultured cells, often by employing fluorescently labeled proteins to allow their detection and visualization (Bates et al. 2015). In conventional fluorescence microscopy, aggregates are typically detected as μm-sized inclusions, often with a roughly spherical appearance. However, advanced imaging techniques such as super-resolution microscopy, cryogenic electron microscopy or electron tomography have revealed that these large inclusions contain narrow filamentous aggregates with widths closer to the nm scale (Duim et al. 2014, Bäuerlein et al. 2017, Peskett et al. 2018, Galaz-Montoya et al. 2021). Notably, such narrow filaments would have been undetectable (as aggregates) in conventional fluorescence microscopy.

The propensity of polyQ segments (and polyQ-containing proteins) to form filamentous structures has been known since their initial identification in the context of HD (Perutz et al. 1994, Scherzinger et al. 1997, 1999). This is based on *in vitro* studies using purified polypeptides or proteins, which permitted a closer analysis of their structural features by various methods. Atomic Force Microscopy (AFM) and negative stain Transmission Electron Microscopy (TEM) revealed fibrillar aggregates reminiscent of amyloid-like fibrils formed by Aβ, α-synuclein, and other amyloidogenic proteins (Scherzinger et al. 1997, 1999, Dahlgren et al. 2005). Solid evidence of the amyloid-like nature of the polyQ aggregates stems from X-ray diffraction studies that detected the cross-β signature, which is the hallmark of amyloid fibril structure (Figure 2.3a) (Perutz et al. 1994). This signature reflects that fibrils from polyQ peptides or polyQ-containing proteins feature a core structure formed by intermolecular β-sheets, in which numerous (intermolecular) hydrogen bonds stabilize the fibril architecture. In polyQ fibrils, these β-sheets are antiparallel, such that β-strands alternate in their direction along the fibril long axis. Interestingly, this is distinct from most other amyloids (related to diseases), which are more typically formed via parallel β-sheets. Nonetheless, polyQ-driven aggregates are generally considered one of the amyloid fibrils, rendering the polyQ expansion disorders among the amyloid diseases.

A more detailed, atomic-level understanding of the structure of polyQ protein aggregates has remained a challenge. Recent years have seen breakthroughs in the structure of other disease-associated amyloids through cryo-EM and solid-state NMR spectroscopy; however, these techniques are yet to produce a full atomistic structure of polyQ amyloids. However, integrating methods that provide complementary insights has gradually led to a more robust understanding of these elusive amyloid fibrils. Early studies of their structure, which relied on diffraction and scattering techniques, yielded various structural models (Figures 2.3b and c) (Perutz et al. 1994, 2002, Sharma et al. 2005, Sikorski and Atkins 2005). Two general categories of proposed configurations were proposed by Perutz and co-workers: variants of tightly packed polar zipper and tubular structures with a water-filled hole in a β-sheet-based nanotube (Figures 2.3b and c). However, the nanotube model has received less support, with most evidence favoring a polar-zipper-like, antiparallel β-sheet structure. Recent advancements, particularly through solid-state NMR spectroscopy, have offered detailed structural insights into the polyQ aggregates and HTTex1 fibrils, with the latter relevant to HD, as noted above (Matlahov and Van der Wel 2019, Van der Wel 2024). It is important to note, however, that these NMR data excel at revealing local atom-level structure and dynamics on the sub-nm scale but that they are best combined with other techniques that provide complementary data. For instance, the combination of Electron Spin Resonance (ESR) spectroscopy, infrared and Raman spectroscopy, electron microscopy, AFM, and other structural techniques have yielded valuable insights. Indeed, the study of polyQ protein aggregates exemplifies the power of integrative and interdisciplinary research.

We will not provide a detailed history of polyQ structural studies, referring readers to already mentioned papers and relevant review articles (Matlahov and Van der Wel 2019, Van der Wel 2024). Here, we focus on a brief description of our current understanding of the structure of polyQ aggregates. As noted above, polyQ amyloid structures are characterized by antiparallel β-sheets assembled from hundreds of proteins per fibril. Unlike other amyloids, which often consist of a small and defined number of proteins per fibril cross-section layer (e.g., two or three proteins), polyQ fibrils are wider and adopt a block-like architecture that is nanometers wide. This is, for instance, visible in the cryo-EM data of HTTex1 fibrils (Figure 2.3d) (Nazarov et al. 2022). An important consequence of this block-like architecture is that a much smaller fraction of the polyQ segments remain accessible on the fibril surface, with a majority buried in the fibril core. This is distinct from most other amyloid fibrils, such as those illustrated in Figure 2.3g (Iadanza et al. 2018). Within the fibril core, the polyQ segments are buried in a highly ordered dehydrated structure, stabilized by hydrogen bonding between protein backbones and side chains. This organization follows the principles of the polar zipper motif proposed by Perutz and co-workers (Perutz et al. 1994). The extended side chains within the β-sheets engage in hydrogen bonding, while a steric-zipper interaction occurs at the interfaces between sheets (Figures 2.3e and f). This structural motif had been suggested in various early studies, and found explicit support from structural solid-state NMR measurements that probed the conformations of the Gln side chains (Hoop et al. 2016). Spectroscopic analysis of both polyQ peptide fibrils and HTTex1 fibrils revealed an interesting aspect of the non-local tertiary structure of the polyQ chain in the fibril core (Buchanan et al. 2014, Hoop et al. 2016). These analyses, which involved fibrils assembled from differently labeled

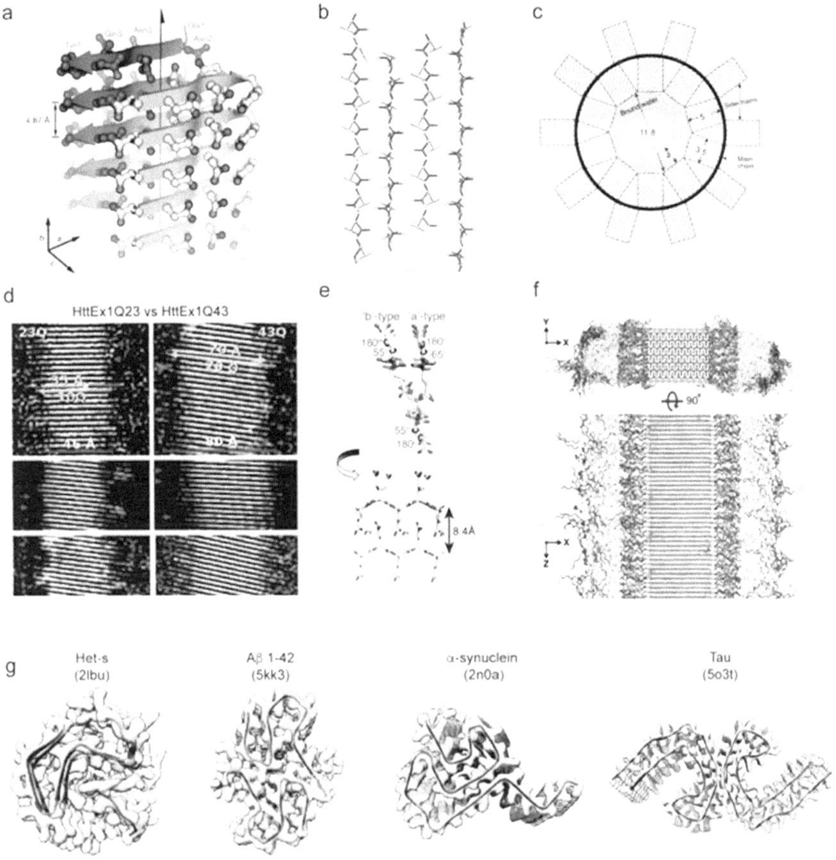

Figure 2.3 Mutated huntingtin structure and other polyQ diseases. (a) Representation of a canonical cross-β structure in amyloid-like crystals of the GNNQQNY peptide. Adapted from ref. (Nelson et al. 2005) with permission. (b) Model of the polar zipper proposed by Perutz and co-workers for polyQ aggregates. Adapted with permission from (Perutz et al. 1994). (c) Schematic showing the tubular polyQ structure, also proposed by Perutz and co-workers. Adapted with permission from (Perutz et al. 1994, 2002). (d) 2D class average images of HTTex1-23Q (left panel) and HTTex1-43Q (right panel) fibrils, showing the order polyQ fibril core. Adapted with permission from (Nazarov et al. 2022). (e) Schematic illustration of the polyQ core structure informed by solid-state NMR structural analysis. Adapted from (Hoop et al. 2016). (f) Atomic-level structural model of mutant HttEx1 fibril, showing a cross-section (top) and side view (bottom) of the fibrils. Both the ordered polyQ core and more flexible and disordered flanking segments (on the surface) are visible. Reproduced with permission from (Bagherpoor Helabad et al. 2024). (g) Representations of near-atomic-resolution models of several other different amyloid fibrils, showing the qualitatively different assembly of the HTTex1 fibril structure. Adapted from (Iadanza et al. 2018) with permission.

protein monomers, showed clearly that there are short-range intra-protein interactions in addition to the typical inter-molecular hydrogen bonding seen in other amyloids. Thus, these data revealed that polyQ fibrils (at least for long polyQ segments) can feature a β-hairpin structure within the polyQ stretch. Furthermore, according to the solid-state NMR data, the core consists of two alternating glutamine conformers, known as "a" and "b" (Figure 2.3f), as a consequence of the antiparallel β-sheet structure (Schneider et al. 2011, Hoop et al. 2016, Bagherpoor Helabad et al. 2024).

Although not the focus of this chapter, the non-polyQ segments and their fate in the aggregates are highly relevant. The abovementioned N17 and PRD segments strongly impact the aggregation behavior of polyQ proteins *in vitro*, and in cellular studies, they also change the associated toxicity of the polyQ or HTTex1 protein (Bates et al. 2015). Remarkably, in the polyQ protein aggregates studied so far, these so-called 'flanking segments' do not appear to be incorporated into the amyloid architecture formed by the polyQ stretch. Instead, they remain outside the core, lacking the β-sheet structure, and retain a high degree of flexibility (Bugg et al. 2012, Lin et al. 2017, Falk et al. 2020). Indeed, they form a 'fuzzy coat' on the fibrils and are barely visible in the EM (Figure 2.3d). While such fuzzy coats are now recognized as common and functionally important among amyloid fibrils (Bhopatkar and Kayed 2023). However, the polyQ case stands out by the seemingly defined boundary between the fibril core and fuzzy coat (Figure 2.3f). This boundary is observed to align closely with the boundaries of the polyQ segment. In other amyloid proteins, the need to align specific residues in subsequent proteins ensures a well-defined fibril core. In contrast, the repetitive nature of the polyQ primary sequence permits for differing alignments between subsequent monomers (without large energetic penalties). This phenomenon leads to an increased disorder in polyQ fibrils, resulting in a more irregular and uneven fibril surface compared to other amyloids.

POLYMORPHISM OF POLYQ PROTEIN AGGREGATES

Perhaps not surprisingly, changing the polyQ length alters the fibrils' appearance and width. In addition, changes in the sequences surrounding the polyQ core also affect the fibril morphology (as well as their aggregation propensity or kinetics). An important consequence of such changes is that they can also affect the biological properties of the (polyQ protein) aggregates. Such a sequence can modulate the (mutant) proteins' toxic effects on (neuronal) cells, as discussed below. But, also the exact same protein (without sequence changes) displays polymorphism of its aggregates. A common feature of amyloid fibrils (and other protein aggregates) is that a particular protein or polypeptide can form a diverse range of structurally distinct aggregates. For instance, the Aβ1-42 peptide can form quite distinct fibril forms, in which the core β-sheet structure and its higher-order assembly can vary (one of these is shown in Figure 2.3g). Also, the HTTex1 aggregates are prone to polymorphism, observable through a variety of structural analysis methods. One of the most obvious changes is that the same HTTex1 variant (with a particular polyQ length) can form fibrils with different widths (Lin et al. 2017, Boatz et al. 2020, Isas et al. 2021). It is important to stress that these comments refer to aggregates formed from purified protein (not in cells). Thanks to advances in structural biology methods and in particular cryogenic Electron Tomography (cryo-ET), we are also gaining a better understanding of the structure of HTTex1 aggregates formed in a cellular context (Bäuerlein et al. 2017, Peskett et al. 2018). One thing such studies reveals that they show the presence of numerous nm-scale (in width) fibrils within μm-sized inclusions seen in cells. Such observations are also supported by several super-resolution microscopy studies that highlight the co-existence of 'large' inclusions with nm-scale (isolated) fibrils (Duim et al. 2014). In summary, these HTTex1 polyQ protein aggregates are complex, heterogeneous, diverse in structure, and are being studied in a variety of contexts.

MECHANISTIC ASPECTS OF POLYQ PROTEIN AGGREGATION

The polyQ-driven aggregation process appears to occur via a nucleated growth mechanism (Bhattacharyya et al. 2005, Kar et al. 2013). Nucleated growth implies that a bottleneck in the aggregation process is the formation of a critical nucleus. This nucleus typically represents an abnormally folded cluster of proteins (small oligomer) that initiates a down-hill aggregation process. *In vitro* studies have provided valuable insights into the molecular mechanisms. A key step in this process is the formation of β-sheet structure within the polyQ segment, which establishes the initial self-propagating structure. Multiple pieces of experimental evidence pinpoint the formation of β-hairpin structures within the expanded polyQ as a pivotal nucleating step accessible only to mutant proteins (Chen et al. 2002, Kar et al. 2011). For polyQ peptides, the size of the critical nucleus varies with the length of the glutamine tract. When the polyQ is short, multiple peptides have to come together to form a good nucleating structure. In contrast, a longer polyQ region enables the formation of a β-hairpin, leading to faster nucleation and faster aggregation. Concentration is another critical parameter, with studies showing concentration-dependent aggregation. Higher concentrations facilitate a speed-up of the aggregation process (Kar et al. 2011). Additionally, the presence of flanking segments, on either side of the polyQ, influence the aggregation mechanism and kinetics. This explains the differences between the polyQ disorders, as each of them has the polyQ flanked by different sequences. In HTTex1, the two flanking segments lack folded domains. Nonetheless, they modulate the kinetics in dramatic and opposing ways. The N17 interacts with other N17 segments and thus can initiate and accelerate HTTex1 aggregation (Thakur et al. 2009). Conversely, the PRD reduces the aggregation propensity of the preceding polyQ segment. The resulting HTTex1 aggregation process is complex and multifaceted, with multiple parallel aggregation pathways that can frustrate attempt to design inhibitors (Jayaraman et al. 2012).

RELEVANCE TO DISEASE MECHANISMS

In the above, we have mainly focused on mechanistic and structural details that could only be elucidated by careful *'in vitro'* experiments with purified proteins under controlled conditions. Naturally, an important question is whether and how we can relate these findings to clinically relevant information. A central question in HD research asks about the processes (or molecules) that are (directly) responsible for the degradation of brain cells, as observed in post-mortem analysis. Given the monogenic nature of this disorder, the mutated HTT gene (or protein) is an obvious candidate. The consensus is that a loss of HTT function (due to the mutation) is not the primary cause of disease, instead focusing on a gain-of-function mechanism. The latter mechanism is based on the idea that the mutant protein (or mutant gene on the RNA or DNA level) gains a new function or property (absent in wild-type counterpart), rendering it 'toxic' to cells. For instance, exposing cultured neuronal cells to aggregated polyQ proteins is observed to reduce their viability and induce cell death (Yang et al. 2002, Nekooki-Machida et al. 2009, Kar et al. 2014, Lin et al. 2017, Isas et al. 2021). Thus, features of aggregated and misfolded proteins appear to be disruptive to cellular processes to the point of toxic consequences for the affected cells. The idea that aggregates could be pathogenic has been part of the HD literature ever since the discovery of protein inclusions in post-mortem tissue. However, some research argues against a direct and straightforward correlation between measurable 'toxicity' and observable aggregates (Arrasate et al. 2004). Indeed, these experiments even seem to suggest that the amount of visible aggregates shows a reverse correlation to toxicity, such that the large inclusions may be protective. Thus, at this time the role of aggregates in HD toxicity remains under debate in HD literature as it is in other amyloid diseases (Caughey and Lansbury 2003, Arrasate and

Finkbeiner 2012). Other candidates for toxic players include misfolded oligomers and monomers of mutant HTT (or its fragments) (Nagai et al. 2007, Takahashi et al. 2008, Hoffner and Djian 2014). In other words, this area continues to be under active investigation. One interesting aspect of the diversity of protein-based mechanisms is that they generally involve a gain of toxic function not by the full-length protein but rather from fragments of the mutant HTT protein. In particular, the already mentioned HTTex1 fragment is seen as the most pathogenic or toxic (Neueder et al. 2017). As noted above, the mutant HTT is prone to the formation of fragments like HTTex1 due to multiple mechanisms, including proteolysis and aberrant splicing.

Given the focus of this chapter and this book, we will briefly discuss some key concepts and ideas relevant to the hypothesized role of aggregated polyQ proteins. One of the ideas that is special to polyQ proteins (and less relevant for non-polyQ amyloids) is that the cell contains a variety of proteins with polyQ segments. It has been suggested that the aggregation of an expanded polyQ protein could lead to the co-aggregation of these other polyQ proteins. Most HD patients are heterozygous, possessing one mutant and one normal HTT allele, and the aggregation of the former likely also encourages the (co)aggregation of the latter. This mechanism is known as (polyQ-based) recruitment or sequestration (Preisinger et al. 1999, Haacke et al. 2006). Moreover, the presence of protein aggregates (in cells) is known to recruit or sequester cellular proteins, as well as proteins lacking polyQ segments. Analysis of polyQ protein aggregates from cells or animal models has identified numerous other proteins captured within, indicating that recruitment is not dependent on polyQ-mediated interactions (Radwan et al. 2017, Juenemann et al. 2018, Riguet et al. 2021). Also, this process could lead to a dysregulation of cellular processes, which may manifest as cytotoxicity. Another common theme in protein misfolding diseases and in HD is that the oligomerizing or aggregating proteins block or disrupt the transport between cellular compartments. In the case of mutant HTT, it has been shown to impact on mitochondrial transport and the nuclear pore complex (Grima et al. 2017, Yablonska et al. 2019). Even this non-comprehensive overview illustrates the diversity of possible parallel mechanisms through which protein aggregation impacts cellular function. Here, we remind the reader that the aggregates in HD are found to form intracellularly, in contrast to the plaques seen in AD.

DIAGNOSTICS AND TREATMENTS

A final topic in this chapter concerns the clinical implications of our growing understanding and ongoing research into the unusual processes behind polyQ protein aggregation. The presence of aggregated mutant proteins is recognized as a hallmark of HD and thus is also an important potential biomarker. As such, efforts are made to detect and quantify these inclusions using different approaches. Positron Emission Tomography (PET) is an important clinical tool in AD to detect Aβ aggregates in the brains of patients (Cohen et al. 2012). Similarly, efforts are underway to develop PET ligands that are specific for mutant HTT aggregates (Liu et al. 2020, Bertoglio et al. 2022). Another promising approach leverages the ability of polyQ aggregate to self-seed or -propagate to develop highly sensitive assays applicable to body fluids (Ast et al. 2018).

Regarding treatments, a major current effort is reducing the production of the aggregation-prone mutant protein – so-called protein lowering strategies (Tabrizi et al. 2019). Given that protein aggregation kinetics are highly sensitive to protein concentration, lowering the protein level should also reduce problems caused by protein aggregation. A few other relevant approaches are being explored: inhibition of protein cleavage or aberrant splicing, upregulation of protein homeostasis (that would counteract aggregation), or direct interference with the aggregation process. The latter could involve small molecule inhibitors of the aggregation process. For instance, it seems that the best approach to treatment in the end may require a combination of different strategies, e.g., combining well-calibrated protein lowering with downstream modulation of protein misfolding or aggregation.

CONCLUDING REMARKS

This chapter has provided a basic introduction to polyQ protein aggregation and its role in CAG repeat disorders. Due to space limitations, we have focused primarily on HD and omitted numerous details, many of which are highly relevant and interesting. Nonetheless, we hope that some of the unusual and intriguing molecular features can be appreciated from the current chapter, along with a basic understanding of their relevance to clinical aspects of the diseases. It is likely also clear to the reader that much work remains to be done, for instance, in the characterization of other polyQ protein disorders. One important reason for focusing here on HD is that we know a lot less about the other diseases in the family of CAG repeat expansion disorders. We would anticipate that a better understanding of their mechanisms would greatly help efforts to treat the entire family of polyQ disorders, including HD. Moreover, lessons learned in polyQ disease research can be expected to boost our understanding of protein misfolding diseases in general.

REFERENCES

Akçimen, F., S. Martins, C. Liao, C.V. Bourassa, H. Catoire, G.A. Nicholson, et al. 2020. Genome-wide association study identifies genetic factors that modify age at onset in Machado-Joseph disease. Aging. 12: 4742–4756.

Arrasate, M. and S. Finkbeiner. 2012. Protein aggregates in Huntington's disease. Exp. Neurol. 238: 1–11.

Arrasate, M., S. Mitra, E.S. Schweitzer, M.R. Segal and S. Finkbeiner. 2004. Inclusion body formation reduces levels of mutant huntingtin and the risk of neuronal death. Nature. 431: 805–810.

Ast, A., A. Buntru, F. Schindler, R. Hasenkopf, A. Schulz, L. Brusendorf, et al. 2018. mHTT Seeding Activity: A Marker of Disease Progression and Neurotoxicity in Models of Huntington's Disease. Mol. Cell. 71: 675–688.e6.

Atwal, R.S., J. Xia, D. Pinchev, J. Taylor, R.M. Epand and R. Truant. 2007. Huntingtin has a membrane association signal that can modulate huntingtin aggregation, nuclear entry and toxicity. Hum. Mol. Genet. 16: 2600–2615.

Bagherpoor Helabad, M., I. Matlahov, R. Kumar, J.O. Daldrop, G. Jain, M. Weingarth, et al. 2024. Integrative determination of atomic structure of mutant huntingtin exon 1 fibrils implicated in Huntington disease. Nat. Commun. 15: 10793.

Bates, G.P., R. Dorsey, J.F. Gusella, M.R. Hayden, C. Kay, B.R. Leavitt, et al. 2015. Huntington disease. Nat. Rev. Dis. Primers. 1: 1–21.

Bäuerlein, F.J.B., I. Saha, A. Mishra, M. Kalemanov, A. Martínez-Sánchez, R. Klein, et al. 2017. *In situ* Architecture and Cellular Interactions of PolyQ Inclusions. Cell. 171: 179–187.e10.

Bertoglio, D., J. Bard, M. Hessmann, L. Liu, A. Gärtner, S. De Lombaerde, et al. 2022. Development of a ligand for *in vivo* imaging of mutant huntingtin in Huntington's disease. Sci. Transl. Med. 14: eabm3682.

Bhattacharyya, A.M., A.K. Thakur and R. Wetzel. 2005. Polyglutamine aggregation nucleation: Thermodynamics of a highly unfavorable protein folding reaction. Proceedings of the National Academy of Sciences. 102: 15400–15405.

Bhopatkar, A.A. and R. Kayed. 2023. Flanking regions, amyloid cores, and polymorphism: the potential interplay underlying structural diversity. J. Biol. Chem. 299: 1–20.

Boatz, J.C., T. Piretra, A. Lasorsa, I. Matlahov, J.F. Conway and P.C.A. van der Wel. 2020. Protofilament structure and supramolecular polymorphism of aggregated mutant huntingtin exon 1. J. Mol. Biol. 432: 4722–4744.

Buchanan, L.E., J.K. Carr, A.M. Fluitt, A.J. Hoganson, S.D. Moran, J.J. de Pablo, et al. 2014. Structural motif of polyglutamine amyloid fibrils discerned with mixed-isotope infrared spectroscopy. Proc. Natl. Acad. Sci. USA. 111: 5796–5801.

Bugg, C.W., J.M. Isas, T. Fischer, P.H. Patterson and R. Langen. 2012. Structural features and domain organization of huntingtin fibrils. J. Biol. Chem. 287: 31739–31746.

Caughey, B. and P.T. Lansbury. 2003. Protofibrils, pores, fibrils, and neurodegeneration: separating the responsible protein aggregates from the innocent bystanders. Annu. Rev. Neurosci. 26: 267–298.

Chen, S., V. Berthelier, W. Yang and R. Wetzel. 2001. Polyglutamine aggregation behavior in vitro supports a recruitment mechanism of cytotoxicity. J. Mol. Biol. 311: 173–182.

Chen, S., F.A. Ferrone and R. Wetzel. 2002. Huntington's disease age-of-onset linked to polyglutamine aggregation nucleation. Proceedings of the National Academy of Sciences of the United States of America. 99: 11884–11889.

Chiti, F. and C.M. Dobson. 2017. Protein Misfolding, Amyloid Formation, and Human Disease: A Summary of Progress Over the Last Decade. Annu. Rev. Biochem. 86: 27–68.

Cohen, A.D., G.D. Rabinovici, C.A. Mathis, W.J. Jagust, W.E. Klunk and M.D. Ikonomovic. 2012. Using pittsburgh compound B for *in vivo* PET imaging of fibrillar amyloid-beta. Adv. Pharmacol. 64: 27–81.

Dahlgren, P.R., M.A. Karymov, J. Bankston, T. Holden, P. Thumfort, V.M. Ingram, et al. 2005. Atomic force microscopy analysis of the huntington protein nanofibril formation. Nanomed. 1: 52–57.

DiFiglia, M., E. Sapp, K.O. Chase, S.W. Davies, G.P. Bates, J.P. Vonsattel, et al. 1997. Aggregation of huntingtin in neuronal intranuclear inclusions and dystrophic neurites in brain. Science. 277: 1990–1993.

Duim, W.C., Y. Jiang, K. Shen, J. Frydman and W.E. Moerner. 2014. Super-resolution fluorescence of huntingtin reveals growth of globular species into short fibers and coexistence of distinct aggregates. ACS Chem. Biol. 9: 2767–2778.

Falk, A.S., J.M. Bravo-Arredondo, J. Varkey, S. Pacheco, R. Langen and A.B. Siemer. 2020. Structural model of the Proline-Rich Domain of Huntingtin Exon-1 fibrils. Biophys. J. 119: 2019–2028.

Galaz-Montoya, J.G., S.H. Shahmoradian, K. Shen, J. Frydman and W. Chiu. 2021. Cryo-electron tomography provides topological insights into mutant huntingtin exon 1 and polyQ aggregates. Commun. Biol. 4: 1–9.

Grima, J.C., J.G. Daigle, N. Arbez, K.C. Cunningham, K. Zhang, J. Ochaba, et al. 2017. Mutant huntingtin disrupts the nuclear pore complex. Neuron. 94: 93–107.e6.

Haacke, A., S.A. Broadley, R. Boteva, N. Tzvetkov, F.U. Hartl and P. Breuer. 2006. Proteolytic cleavage of polyglutamine-expanded ataxin-3 is critical for aggregation and sequestration of non-expanded ataxin-3. Hum. Mol. Genet. 15: 555–568.

Harding, R.J., P. Loppnau, S. Ackloo, A. Lemak, A. Hutchinson, B. Hunt, et al. 2019. Design and characterization of mutant and wildtype huntingtin proteins produced from a toolkit of scalable eukaryotic expression systems. J. Biol. Chem. 294: 6986–7001.

Hoffner, G. and P. Djian. 2014. Monomeric, oligomeric and polymeric proteins in Huntington disease and other diseases of polyglutamine expansion. Brain Sci. 4: 91–122.

Hoop, C.L., H.-K. Lin, K. Kar, G. Magyarfalvi, J.M. Lamley, J.C. Boatz, et al. 2016. Huntingtin exon 1 fibrils feature an interdigitated β-hairpin-based polyglutamine core. Proc. Natl. Acad. Sci. USA. 113: 1546–1551.

Hoschek, F., J. Natan, M. Wagner, K. Sathasivam, A. Abdelmoez, B. Von Einem, et al. 2024. Huntingtin HTT1a is generated in a CAG repeat-length-dependent manner in human tissues. Mol. Med. 30: 1–11.

Iadanza, M.G., M.P. Jackson, E.W. Hewitt, N.A. Ranson and S.E. Radford. 2018. A new era for understanding amyloid structures and disease. Nat. Rev. Mol. Cell Biol. 19: 755–773.

Isas, J.M., N.K. Pandey, H. Xu, K. Teranishi, A.K. Okada, E.K. Fultz, et al. 2021. Huntingtin fibrils with different toxicity, structure, and seeding potential can be interconverted. Nat. Commun. 12: 1–11.

Jayaraman, M., R. Kodali and R. Wetzel. 2009. The impact of ataxin-1-like histidine insertions on polyglutamine aggregation. Protein Eng. Des. Sel. 22: 469–478.

Jayaraman, M., R. Mishra, R. Kodali, A.K. Thakur, L.M.I. Koharudin, A.M. Gronenborn, et al. 2012. Kinetically competing huntingtin aggregation pathways control amyloid polymorphism and properties. Biochemistry. 51: 2706–2716.

Johnson, S.L., W.-L. Tsou, M.V. Prifti, A.L. Harris and S.V. Todi. 2022. A survey of protein interactions and posttranslational modifications that influence the polyglutamine diseases. Front. Mol. Neurosci. 15: 974167.

Juenemann, K., A.H.P. Jansen, L. van Riel, R. Merkx, M.P.C. Mulder, H. An, et al. 2018. Dynamic recruitment of ubiquitin to mutant huntingtin inclusion bodies. Sci. Rep. 8: 1–15.

Kar, K., M. Jayaraman, B. Sahoo, R. Kodali and R. Wetzel. 2011. Critical nucleus size for disease-related polyglutamine aggregation is repeat-length dependent. Nature Struct. Mol. Biol. 18: 328–336.

Kar, K., C.L. Hoop, K.W. Drombosky, M.A. Baker, R. Kodali, I. Arduini, et al. 2013. β-hairpin-mediated nucleation of polyglutamine amyloid formation. J. Mol. Biol. 425: 1183–1197.

Kar, K., I. Arduini, K.W. Drombosky, P.C.A. van der Wel and R. Wetzel. 2014. D-polyglutamine amyloid recruits L-polyglutamine monomers and kills cells. J. Mol. Biol. 426: 816–829.

Kennedy, L. 2003. Dramatic tissue-specific mutation length increases are an early molecular event in Huntington disease pathogenesis. Hum. Mol. Genetics. 12: 3359–3367.

Kuiper, E.F.E., E.P. De Mattos, L.B. Jardim, H.H. Kampinga and S. Bergink. 2017. Chaperones in polyglutamine aggregation: Beyond the Q-stretch. Front. Neurosci. 11: 1–11.

Landles, C. and G.P. Bates. 2004. Huntingtin and the molecular pathogenesis of Huntington's disease. EMBO Reports. 5: 958–963.

Lin, H.-K., J.C. Boatz, I.E. Krabbendam, R. Kodali, Z. Hou, R. Wetzel, et al. 2017. Fibril polymorphism affects immobilized non-amyloid flanking domains of huntingtin exon1 rather than its polyglutamine core. Nat. Commun. 8: 1–12.

Liu, Y. and S.H. Wilson. 2012. DNA base excision repair: a mechanism of trinucleotide repeat expansion. Trends Biochem. Sci. 37: 162–172.

Liu, L., M.E. Prime, M.R. Lee, V. Khetarpal, C.J. Brown, P.D. Johnson, et al. 2020. Imaging mutant huntingtin aggregates: Development of a potential PET ligand. J. Med. Chem. 63: 8608–8633.

Macdonald, M. 1993. A novel gene containing a trinucleotide repeat that is expanded and unstable on Huntington's disease chromosomes. Cell. 72: 971–983.

Matlahov, I. and P.C.A. van der Wel. 2019. Conformational studies of pathogenic expanded polyglutamine protein deposits from Huntington's disease. Exp. Biol. Med. 244: 1584–1595.

McMurray, C.T. 2010. Mechanisms of trinucleotide repeat instability during human development. Nat. Reviews. Genetics. 11: 786–799.

Miguez, A., C. Gomis, C. Vila, M. Monguió-Tortajada, S. Fernández-García, G. Bombau, et al. 2023. Soluble mutant huntingtin drives early human pathogenesis in Huntington's disease. Cell. Mol. Life. Sci. 80: 238.

Mouro Pinto, R., L. Arning, J.V. Giordano, P. Razghandi, M.A. Andrew, T. Gillis, et al. 2020. Patterns of CAG repeat instability in the central nervous system and periphery in Huntington's disease and in spinocerebellar ataxia type 1. Hum. Mol. Genet. 29: 2551–2567.

Nagai, Y., T. Inui, H.A. Popiel, N. Fujikake, K. Hasegawa, Y. Urade, et al. 2007. A toxic monomeric conformer of the polyglutamine protein. Nat. Struct. Mol. Biol. 14: 332–340.

Nazarov, S., A. Chiki, D. Boudeffa and H.A. Lashuel. 2022. Structural basis of huntingtin fibril polymorphism revealed by cryogenic electron microscopy of exon 1 HTT fibrils. J. Am. Chem. Soc. 144: 10723–10735.

Nekooki-Machida, Y., M. Kurosawa, N. Nukina, K. Ito, T. Oda and M. Tanaka. 2009. Distinct conformations of *in vitro* and *in vivo* amyloids of huntingtin-exon1 show different cytotoxicity. Proc. Natl. Acad. Sci. USA. 106: 9679–9684.

Nelson, R., M.R. Sawaya, M. Balbirnie, A.Ø. Madsen, C. Riekel, R. Grothe, et al. 2005. Structure of the cross-β spine of amyloid-like fibrils. Nature. 435: 773–778.

Neueder, A., C. Landles, R. Ghosh, D. Howland, R.H. Myers, R.L.M. Faull, et al. 2017. The pathogenic exon 1 HTT protein is produced by incomplete splicing in Huntington's disease patients. Sci. Rep. 7: 1–10.

Newcombe, E.A., K.M. Ruff, A. Sethi, A.R. Ormsby, Y.M. Ramdzan, A. Fox, et al. 2018. Tadpole-like conformations of huntingtin exon 1 are characterized by conformational heterogeneity that persists regardless of polyglutamine length. J. Mol. Biol. 430: 1442–1458.

Parsons, M.P. and L.A. Raymond. 2015. Huntington disease. pp. 275-292. *In*: M.J. Zigmond, C.A. Wiley, M.F. Chesselet (eds). Neurobiology of Brain Disorders: Biological Basis of Neurological and Psychiatric Disorders, 2nd Ed. Elsevier.

Perutz, M.F., J.T. Finch, J. Berriman and A. Lesk. 2002. Amyloid fibers are water-filled nanotubes. Proceedings of the National Academy of Sciences. 99: 5591–5595.

Perutz, M.F., T. Johnson, M. Suzuki and J.T. Finch. 1994. Glutamine repeats as polar zippers: their possible role in inherited neurodegenerative diseases. Proceedings of the National Academy of Sciences of the United States of America. 91: 5355–5358.

Peskett, T.R., F. Rau, J. O'Driscoll, R. Patani, A.R. Lowe and H.R. Saibil. 2018. A liquid to solid phase transition underlying pathological huntingtin exon1 aggregation. Mol. Cell. 70: 588–601.e6.

Preisinger, E., B.M. Jordan, A. Kazantsev and D. Housman. 1999. Evidence for a recruitment and sequestration mechanism in Huntington's disease. Philos. Trans. R. Soc. Lond., B., Biol. Sci. 354: 1029–1034.

Radwan, M., R.J. Wood, X. Sui and D.M. Hatters. 2017. When proteostasis goes bad: Protein aggregation in the cell. IUBMB Life. 69: 49–54.

Reiner, A., I. Dragatsis and P. Dietrich. 2011. Genetics and neuropathology of Huntington's disease. Int. Rev. Neurobiol. 98: 325–372.

Riguet, N., A.-L. Mahul-Mellier, N. Maharjan, J. Burtscher, M. Croisier, G. Knott, et al. 2021. Nuclear and cytoplasmic huntingtin inclusions exhibit distinct biochemical composition, interactome and ultrastructural properties. Nat. Commu. 12: 1–27.

Saudou, F. and S. Humbert. 2016. The biology of huntingtin. Neuron. 89: 910–926.

Schaefer, M.H., E.E. Wanker and M.A. Andrade-Navarro. 2012. Evolution and function of CAG/polyglutamine repeats in protein–protein interaction networks. Nucleic Acids Res. 40: 4273–4287.

Scherzinger, E., R. Lurz, M. Turmaine, L. Mangiarini, B. Hollenbach, R. Hasenbank, et al. 1997. Huntingtin-encoded polyglutamine expansions form amyloid-like protein aggregates *in vitro* and *in vivo*. Cell. 90: 549–558.

Scherzinger, E., A. Sittler, K. Schweiger, V. Heiser, R. Lurz, R. Hasenbank, et al. 1999. Self-assembly of polyglutamine-containing huntingtin fragments into amyloid-like fibrils: Implications for Huntington's disease pathology. Proceedings of the National Academy of Sciences. 96: 4604–4609.

Schneider, R., M.C. Schumacher, H. Mueller, D. Nand, V. Klaukien, H. Heise, et al. 2011. Structural characterization of polyglutamine fibrils by solid-state NMR spectroscopy. J. Mol. Biol. 412: 121–136.

Schulte, J. and J.T. Littleton. 2011. The biological function of the Huntingtin protein and its relevance to Huntington's disease pathology. Curr. Trends Neurol. 5: 65–78.

Sharma, D., L.M. Shinchuk, H. Inouye, R. Wetzel and D.A. Kirschner. 2005. Polyglutamine homopolymers having 8–45 residues form slablike β-crystallite assemblies. Proteins: Structure, Function, and Bioinformatics. 61: 398–411.

Sikorski, P. and E. Atkins. 2005. New model for crystalline polyglutamine assemblies and their connection with amyloid fibrils. Biomacromolecules. 6: 425–432.

Stevenson, C.S. 1934. A biography of george huntington. M.d. Bulletin of the Institute of the History of Medicine. 2: 53–76.

Tabrizi, S.J., R. Ghosh and B.R. Leavitt. 2019. Huntingtin lowering strategies for disease modification in Huntington's disease. Neuron. 101: 801–819.

Takahashi, T., S. Kikuchi, S. Katada, Y. Nagai, M. Nishizawa and O. Onodera. 2008. Soluble polyglutamine oligomers formed prior to inclusion body formation are cytotoxic. Hum. Mol. Genet. 17: 345–356.

Tao, M., N.K. Pandey, R. Barnes, S. Han and R. Langen. 2019. Structure of membrane-bound huntingtin exon 1 reveals membrane interaction and aggregation mechanisms. Structure. 27: 1570-1580.e4.

Thakur, A.K., M. Jayaraman, R. Mishra, M. Thakur, V.M. Chellgren, I.-J.L. Byeon, et al. 2009. Polyglutamine disruption of the huntingtin exon 1 N terminus triggers a complex aggregation mechanism. Nat. Struct. Mol. Biol. 16: 380–389.

Van der Wel, P.C.A. 2024. Solid-state nuclear magnetic resonance in the structural study of polyglutamine aggregation. Biochem. Soc. Trans. 52: 719–731.

Waldvogel, H.J., Kim, E.H., Tippett, L.J., Vonsattel, JP.G., Faull, R.L. 2015. The Neuropathology of Huntington's disease. pp. 33–80. *In*: H. Nguyen, and M. Cenci (eds). Behavioral Neurobiology of Huntington's Disease and Parkinson's Disease. Current Topics in Behavioral Neurosciences, vol 22. Springer, Berlin, Heidelberg.

Wanker, E.E. 2000. Protein aggregation and pathogenesis of Huntington's disease: Mechanisms and correlations. Biol. Chem. 381: 937–942.

Yablonska, S., V. Ganesan, L.M. Ferrando, J. Kim, A. Pyzel, O.V. Baranova, et al. 2019. Mutant huntingtin disrupts mitochondrial proteostasis by interacting with TIM23. Proceedings of the National Academy of Sciences. 116: 16593–16602.

Yang, W., J.R. Dunlap, R.B. Andrews and R. Wetzel. 2002. Aggregated polyglutamine peptides delivered to nuclei are toxic to mammalian cells. Hum. Mol. Genet. 11: 2905–2917.

Chapter **3**

Importance of Surface Hydrophobic Clusters in Proteins

Fabio C.L. Almeida*[1,2], Ramon Pinheiro-Aguiar[1,2] and Ariana Azevedo Vasconcelos[1,2]

[1]Institute of Medical Biochemistry, Federal University of Rio de Janeiro, Rio de Janeiro, Brazil

[2]National Center of Nuclear Magnetic Resonance (CNRMN), CENABIO, Federal University of Rio de Janeiro, Rio de Janeiro, Brazil

INTRODUCTION

The pioneering work of Anfinsen may be the most important milestone for understanding protein folding (Anfinsen et al. 1961, Anfinsen 1972, 1973). His studies on the renaturation of ribonuclease A (RNase) support the thermodynamics hypothesis, which states that the native state of the protein structure is the one in which the Gibbs free energy of the whole system is the lowest.

One of the main factors that drives protein folding is the hydrophobic effect, which ultimately leads to hydrophobic collapse during protein folding, driving the formation of a hydrophobic core in globular proteins.

*For Correspondence: falmeida@bioqmed.ufrj.br

Kauzmann first proposed that the increase in the entropy of the system upon burial of the hydrophobic side chains drives the protein chain to fold to the native state (Kauzmann 1959). The total entropy of the system increases mainly because of the increase in the solvent entropy.

The reason for the entropic penalty on the solvation of hydrophobic side chains was initially understood as the formation of a water clathrate, which was controversial (Baldwin and Rose 2016). Clathrate is a type of chemical structure in which an apolar hydrocarbon is trapped within a cage-like framework formed by a highly ordered lattice of water. More recent studies have shown that water hydration shells are stabilized by Van De Walls (VDW) interactions between the hydrocarbons of the solute and water. The extensive decrease in the entropy of the solvent results from the ordering of the water molecules in the first layer around the hydrocarbon (Jorgensen et al. 1985). The latest conclusions support the Kauzmann view of the hydrophobic effect and its role in protein folding (Baldwin 2014, Baldwin and Rose 2016). Despite the increase in our comprehension of the thermodynamics of hydrocarbon hydration, "understanding the protein-folding problem is limited by our inability to predict how hydrophobic residues interact with water and with each other to yield free-energy values that can drive the folding process", as stated by Baldwin and Rose (Baldwin and Rose 2016).

More recent studies on the hydration of hydrocarbons have demonstrated that, for most amino acid side chains, hydration entropy is unfavorable (Table 3.1) (Wolfenden et al. 2015). In other words, all side chains, even the polar ones, have at least some hydrophobic nature. The hydration shells of the aliphatic and aromatic moieties of the side chains, which are more prevalent in the hydrophobic cores of globular proteins, exhibit highly unfavorable entropic and enthalpic contributions. Even polar side chains, such as arginine and glutamine, exhibit highly unfavorable entropic contributions. For these polar side chains, particularly those with long chains, their localization on the protein surface is driven by favorable enthalpic contributions associated with hydration. These observations suggest that the hydrophobic effect is not limited to the protein core but is also a significant factor on the protein surface. In this chapter, we describe how plant defensins (Machado et al. 2018, Pinheiro-Aguiar et al. 2020, 2023), disintegrins (Vasconcelos et al. 2021, 2022, 2024), toxins, human defensins, chemokines, and other proteins highly cross-linked with disulfide bonds lack a canonical hydrophobic core and display all hydrophobic side chains exposed to the protein surface. These proteins serve as valuable models for understanding the contribution of hydrophobic effect on protein surfaces.

Table 3.1 Thermodynamics of transfer from Water (W) to Cyclohexane (C) (kcal/mol at 25°C) of molecules mimetic of the amino acid side chains. K_{W-C} is the equilibrium constant for the transfer of hydrocarbons from a neutral aqueous solution to cyclohexane. The $\log K_{W-C}$, ΔG_{W-C}, ΔH_{W-C}, and $-T\Delta S_{W-C}$ data were extracted from Wolfenden (2015) (Wolfenden et al. 2015). The Zhou and Zhou hydrophobicity scale (Zhou and Zhou 2004) quantifies the frequency with which each amino acid is buried in the protein core. This scale refines the statistical approach of the Janin scale (Janin 1979). This parameter was used to infer the likelihood of the residues being buried (+) or at the protein surface (–).

Amino acid	Hydrocarbons	$\text{Log}K_{W-C}$	ΔG_{W-C}	ΔH_{W-C}	$-T\Delta S_{W-C}$	Hydrophobicity (Zhou and Zhou)	Likely localization
Arg	n-Propylguanidine	–4.32	5.88	13.74	–7.86	–1.24	surface
Glu	Propionic acid	–2.26	3.07	10.35	–7.28	–0.7	surface
Val	Propane	4.09	–5.56	1.48	–7.04	1.63	buried
His	4-Methylimidazole	–3.49	4.75	11.59	–6.84	–0.4	surface
Ile	Butane	4.24	–5.77	1.03	–6.8	1.97	buried
Leu	Isobutane	4.24	–5.77	0.6	–6.37	1.32	buried
Pro	Triangulation	3.75	–5.1	0.54	–5.64	–0.39	surface
Ala	Methane	2.11	–2.87	2.55	–5.42	0.33	buried
Cys	Methanethiol	1.53	–2.08	3.12	–5.2	0.71	buried
Lys	n-Butylamine	–0.27	0.37	5.31	–4.94	–1.18	surface
Thr	Ethanol	–1.83	2.49	6.72	–4.23	–0.27	surface
Tyr	p-Cresol	–0.31	0.42	4.35	–3.93	0.02	surface/buried
Asp	Acetic acid	–3.29	4.47	8.34	–3.86	–0.77	surface
Met	Ethylmethylsulfide	1.91	–2.6	0.8	–3.4	0.67	buried
Phe	Toluene	2.64	–3.59	–0.25	–3.34	1.3	buried
Gln	Propionamide	–4.07	5.54	8.79	–3.25	–0.73	surface
Gly	Hydrogen	0.2	–0.27	2.24	–2.52	0.18	surface/buried
Ser	Methanol	–2.82	3.84	6.21	–2.37	–0.37	surface
Trp	3-Methylindole	1.83	–2.49	–0.13	–2.36	0.16	surface/buried
Asn	Acetamide	–4.88	6.64	7.14	–0.51	–0.81	surface

ENERGY LANDSCAPE

The thermodynamics hypothesis led to the development of the energy landscape theory (Bryngelson and Wolynes 1987, Onuchic et al. 1997), which describes a statistical distribution of all thermodynamically accessible conformations. The seminal paper of 1987 introduced the concept of minimal frustration in protein folding, drawing analogies from spin glass theory to explain how proteins fold efficiently despite a complex energy landscape. Studies of the entropy in models of foldable polymers have shown that the conformational ensemble has fewer conformations in the more compact and lowest energy states, taking the form of a funnel-like shape of the energy landscape (Dill and MacCallum 2012). In other words, the funnel-shaped energy landscape shows many high-energy disordered structures and only a few low-energy folded structures. The rugged energy landscape shows hills, kinetic traps, and energy barriers, suggesting that the same protein sequence produces different discrete conformational states. Some will be in the folding pathway, and others will be trapped by high energy barriers. The theory predicts that the conformational states that are in the folding pathway are folding intermediates, which are essential for understanding the general principles of protein folding and how proteins rapidly fold (Dill and MacCallum 2012, Malhotra and Udgaonkar 2016). In a seminal paper, Levinthal pointed to the solution: the protein folds did not experiment with all the available conformational space but was guided by local interactions that were recently recognized as folding intermediates (Levinthal 1968).

The theory is based on the concept of frustrations. In large and complex systems such as proteins, there are always conflicting forces, which can lead to complex structures and behaviors. Frustration refers to situations where conflicting interactions within a protein sequence make it difficult for the protein to find a stable conformation. For example, some interactions might favor one structural arrangement, whereas others might favor another, leading to instability or slow folding. The theory of minimal frustration suggests that, through evolution, proteins have been optimized to reduce these conflicts. The native structure of an ordered protein represents a global minimum in the energy landscape, where most of the interactions between amino acid residues are mutually compatible and stabilize the folded structure.

The theory of minimal frustration is a key concept in understanding protein folding, as it explains how proteins can efficiently fold into their native functional structures despite the vast number of possible conformations (Bryngelson et al. 1995, Wolynes 2004). It was developed to address the Levinthal paradox, in which while proteins could theoretically sample an astronomical number of configurations, they fold rapidly and reproducibly in nature (Levinthal 1968). This theory proposes that the evolution of proteins has shaped their energy landscapes to be "minimally frustrated," allowing them to fold efficiently into their native state (Leopold et al. 1992, Wolynes 2015).

The energy landscape of a minimally frustrated protein is thought to resemble a "funnel." At the top of the funnel, many unfolded conformations with relatively high energy exist. As the protein folds, it moves down the funnel toward the native state, with fewer and fewer conformational choices, as interactions between residues progressively stabilize the structure. In this energy landscape, local minima (representing misfolded or partially folded states) are rare or shallow, allowing the protein to avoid being trapped in nonnative configurations. This funnel-shaped landscape allows proteins to navigate folding pathways efficiently, with minimal barriers to reaching their final structure.

Although the frustration is minimal for the native state of an ordered protein, proteins still exhibit pockets of local frustration, particularly in regions crucial for function, such as active or allosteric sites. These areas may have competing interactions that confer conformational variability, enabling the protein to adjust to environmental shifts or engage with ligands. In this way, while the globular protein general structure is stabilized by reducing frustration, localized frustration introduces the dynamic properties essential for its function (Gianni et al. 2021).

Frustration plays a significant role in the folding and dynamic properties of Intrinsically Disordered Proteins (IDPs), which do not adopt a stable, well-folded structure under physiological conditions (Kulkarni et al. 2022). Instead, IDPs exist as dynamic ensembles of conformations, with their disorder being crucial for their biological function. The inherent flexibility of IDPs allows them to engage in diverse interactions, often involving transient binding and conformational adaptability, which is critical for their roles in processes such as signaling, regulation, and molecular recognition. The balance between local frustration and structural flexibility in IDPs facilitates their ability to respond dynamically to changes in their environment, making them essential components in cellular regulation and signaling pathways (Ferreiro et al. 2014). A large fraction of the proteome is disordered and largely frustrated, comprising 12 to 21% in archaea, 18–28% in bacteria, and increasing to 20–40% in viruses and 35–45% in eukaryotes (Kulkarni et al. 2022).

Therefore, the theory of minimal frustration not only explains efficient protein folding but also emphasizes the finely tuned energetic principles that support both stability and function. The biological roles of proteins are linked to subtle local frustration effects and how this influences the emergence of higher-energy thermally accessible states (excited states), which participate in molecular recognition, catalysis, and allosteric transitions (Whittaker et al. 2011, Stelzl et al. 2020). Nuclear Magnetic Resonance (NMR) is particularly powerful for determining local conformational dynamics (Palmer 2014, Long et al. 2014).

HIGHLY DISULFIDE CROSSLINKED PROTEINS ARE INTRINSICALLY DISORDERED PROTEINS UNDERGOING OXIDATIVE FOLDING

Despite lacking a canonical hydrophobic core, highly disulfide-crosslinked proteins still fold efficiently. Here, we refer to them as coreless proteins, where folding is predominantly governed by surface forces. Anfinsen's seminal work demonstrated that the information required for the folding of disulfide-bonded RNase is encoded in its primary sequence (Anfinsen et al. 1961). These findings showed that oxygen alone was sufficient for oxidative folding, enabling the formation of disulfide bonds. In other words, the primary sequence contains the necessary instructions to bring cysteines into the correct spatial arrangement, with the oxidative stage serving as the next step posttranslational modification (Craik 2011). The formation of disulfide bonds occurs in a hierarchical order in the case of proteins with multiple disulfide bonds (Vinci et al. 2004).

It was later discovered that oxidative folding, in the presence of only oxygen, can take minutes to hours. In cells, this process is catalyzed by a cascade of enzymes that facilitate disulfide bond reshuffling (Kadokura et al. 2003). This reshuffling is critical, allowing the protein to effectively sample the conformational space and achieve a minimally frustrated state. Unlike globular proteins, where folding is driven largely by hydrophobic collapse, coreless proteins rely primarily on surface foldons for the folding process.

The oxidative folding of disulfide bond-containing protein occurs in the lumen of the Endoplasmic Reticulum (ER) in eukaryotic cells. This process is catalyzed mainly by ER sulfhydryl oxidase 1 (Ero1) and Protein Disulfide Isomerase (PDI) a large protein family, having 20 PDIs expressed in humans. Ero1 catalyzes disulfide oxidation, while PDI has the dual function of disulfide isomerization and chaperone activity. Considering the large quantity of proteins that a cell must fold, and the number of possible disulfide isomers in each protein, oxidative protein folding can pose a major challenge to cells. While oxidation is necessary for oxidative protein folding, reduction is equally important for preventing the formation of non-native disulfides and for maintaining ER redox homeostasis. Reduction is conducted by the glutathione and thioredoxin system (Wang and Wang 2023). Oxidation, reduction, and efficient disulfide isomerization warrant the necessary reshuffling to achieve a minimally frustrated state.

In bacteria, the oxidative folding occurs strictly in the periplasm, where the triad oxidation, reduction, and efficient disulfide isomerization is warranted by the transmembrane proteins system DsbB-DsbA and DsbD-DsbC/G (Delaunay-Moisan et al. 2017). Oxidative folding of mitochondrial proteins occurs in the intermembrane space (Berardi et al. 2011, Fraga and Ventura 2012, Zarges and Riemer 2024).

The efficiency and fidelity of disulfide bond formation are not yet completely understood. It is well described the participation and synergistic cooperation of the PDI family, which has their activity regulated by phosphorylation and other post-translational modifications. The reason for the large number of PDIs in higher eukaryotes is most certainly related to their distinct substrate specificity, in which their chaperone activity has a pivotal role, facilitating the folding/unfolding and reshuffling of specific regions of the cysteine-containing protein (Wang and Wang 2023).

Cysteine residues are regarded as "order-promoting" among the amino acids because they can form inter- or intramolecular disulfide bonds, providing significant thermal stability to proteins in oxidizing environments. Although it seems paradoxical, many disulfide-containing proteins have evolved to feature these order-promoting cysteines alongside disorder-promoting sequences. Computational algorithms that predict the degree of disorder in protein sequences, such as IUPred2A and AIUPred (Mészáros et al. 2018, Erdos and Dosztányi 2024) show that cysteine-rich sequences often exhibit pronounced disorder in their reduced form, but not when oxidized, suggesting that these regions may function in a redox-sensitive manner. Numerous protein domains contain disordered regions interspersed with cysteines, and there is a notable positive correlation between the number of cysteines present and the degree of disorder in the surrounding sequences (Bhopatkar et al. 2020a). This tendency of disorder of the disulfide-containing proteins in the reduced state has been shown for several protein domains, such as carboxypeptidase, Kunitz-type protease inhibitor domain, zinc finger motifs, chemokine motifs, defensin folds, three-finger toxin folds, granulin domains (Arolas et al. 2008, Ghag et al. 2016, 2017, Esperante et al. 2017, Bhopatkar et al. 2020a, b, 2022).

IDPs are almost depleted of order-promoting amino acids, such as Trp, Tyr, Phe, Ile, Leu, Val, Cys, and Asn, and are substantially enriched in Ala, as well as in polar, disorder-promoting amino acids, such as Arg, Gly, Gln, Ser, Glu, Lys, and Pro (Bhopatkar et al. 2020a, Kulkarni et al. 2022). Amino acid residues can be ranked by their propensity to promote order or disorder in proteins, with the following sequence (from the most order-promoting to the most disorder-promoting): Cys, Trp, Tyr, Ile, Phe, Val, Leu, His, Thr, Asn, Ala, Gly, Asp, Met, Lys, Arg, Ser, Gln, Pro, and Glu. It is reasonable to assume that the presence of hydrophobic order-promoting amino acids in disulfide-containing sequences has a significant effect on protein oxidative folding by inducing local order and forming the surface hydrophobic clusters in the coreless proteins.

IMPORTANCE OF SURFACE FORCES FOR CORELESS PROTEINS

For coreless proteins, instead of the hydrophobic core present in globular proteins, multiple surface hydrophobic clusters act as independent foldons (Pinheiro-Aguiar et al. 2023). The hierarchy of disulfide bond formation is likely dictated by the relative stability of each surface hydrophobic cluster. As we noted earlier, all amino acid side chains are hydrophobic because the hydration entropy is unfavorable even for the polar side chains (Table 3.1). What drives the side chain to the protein core or to the surface is the enthalpic contribution. The hydration of polar chains is enthalpically favorable, driving them to the protein surface. On the other hand, the hydration of apolar side chains is enthalpically and entropically unfavorable, driving them to the hydrophobic core in globular proteins.

Coreless proteins, such as plant defensins, disintegrins, toxins, and other highly disulfide crosslinked proteins, are highly soluble in water, despite having all the hydrophobic side chains facing the protein surface. The hydrophobic side chains are located on the protein surface but are protected from full exposure to water by adjacent polar side chains, concavities on the protein surface, and/or the shielding effect of protein solvation. This is what makes hydrophobic surface clusters important stabilizing units for coreless proteins and probably for globular proteins as well (Figure 3.1).

PROTEIN SOLVATION

Protein solvation refers to the interaction between protein molecules and the surrounding water and counterions. This process is essential for stabilizing the protein structure, facilitating proper folding, and influencing the dynamics of protein function. Water molecules form hydration shells around the protein, interacting with amino acid residues, particularly those that are polar or charged. These interactions help maintain protein conformation by shielding hydrophobic residues from the aqueous environment and allowing hydrophilic residues to interact with the solvent. This delicate balance of interactions ensures that the protein achieves its correct functional structure.

Research by Arieh Ben Naim noted that the importance of surface forces for protein folding has been underestimated (Ben-Naim 2003, 2012, Durell and Ben-Naim 2017). A possible reason for this is the almost complete lack of direct experimental approaches to study solvation in proteins, especially for hydrophobic side chains.

Tightly bound water solvating the proteins is directly observed in the electron density maps of the protein structure solved by X-ray crystallography, revealing the water molecules that are symmetric in the crystal. This method provides important information on polar interactions, mainly for water molecules that are hydrogen bonded with polar residues at the protein surface (Almeida et al. 2021).

NMR may also provide direct information on protein solvation in solution (Almeida et al. 2021), allowing the observation of more dynamic water molecules. The nuclear spins of water hydrogen nuclei are dipolar-coupled with protein hydrogens. This is achieved by measuring the Nuclear Overhauser Effect (NOE) between water and protein nuclei (Otting et al. 1991, 1997). The challenge is that thousands of nonlocal water molecules contribute to NOE buildup (Modig et al. 2004); thus, studies of solvation in bulk water are nonlocal and limited to protein cavities where water exchange with bulk water is restricted (Cruzeiro-Silva et al. 2014, Iqbal et al. 2015).

A way to avoid the problem of nonlocal water molecules in NOE buildup is to encapsulate the protein in reversed micelles. In this environment, the water pool can be tuned to a minimal amount, just enough to solvate the protein. This method enables the observation of NOEs only between protein and local water molecules (Nucci et al. 2011).

Both X-ray crystallography and NMR can observe only local water molecules with residence times longer than the nanosecond timescale. Qvist and Halle used ^{2}H NMR relaxation to characterize the rotational dynamics of water in hydration shells of hydrophobic solutes (Qvist and Halle 2008). The authors measured the rotational correlation time of bulk water and water in the hydration shell over a wide range of temperatures and reported that water solvating hydrophobic chains rotate with a lower activation energy than bulk water does. Water molecules that solvate hydrophobic residues via van der Waals interactions have lifetimes in the picosecond range. This explains why NMR cannot observe water-solvating hydrophobic side chains of proteins (Baldwin and Rose 2016).

Furthermore, solvation influences protein dynamics by affecting intramolecular interactions and conformational flexibility. Water molecules can penetrate the protein surface, particularly in regions that undergo conformational changes, influencing local mobility and function.

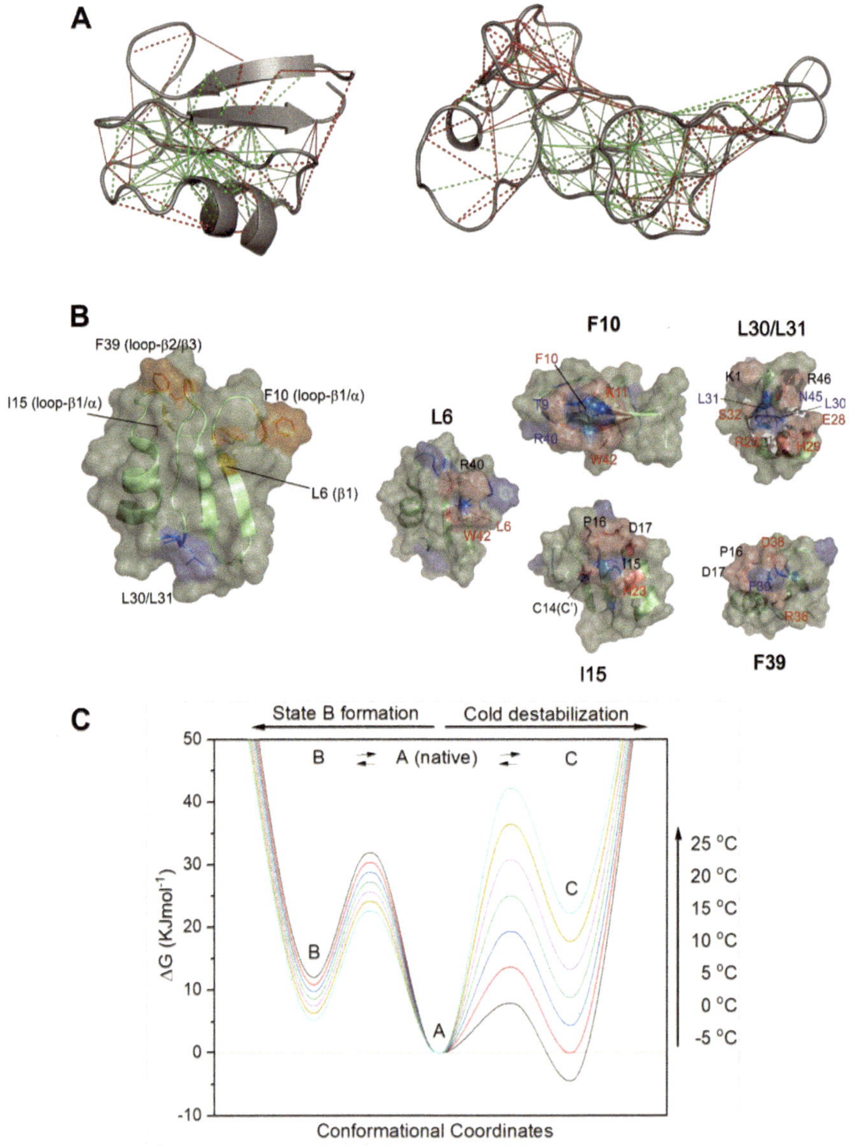

Figure 3.1 Dynamics and frustration of coreless proteins. (A) Mutational frustration of Psd2 (left) and Jarastatin (right) calculated via the Frustratometer server (http://frustratometer.qb.fcen.uba.ar) (Parra et al. 2016). Highly frustrated contacts are in red, and minimally frustrated contacts are in green. Note that almost all frustrated contacts are located at the protein surface. (B) Surface hydrophobic cores present in Psd2. On the left is the localization of hydrophobic exposed residues in the protein. Hydrophobic clusters with neighboring residues

In proteins that experience allosteric regulation, changes in solvation can propagate through the structure, modulating protein activity at sites distant from the solvent-exposed regions. These solvation dynamics are also important in pathological conditions, where disruptions in hydration can lead to protein misfolding and aggregation, as observed in diseases such as Alzheimer's disease and Parkinson's disease. Therefore, the study of protein solvation is central to understanding both the normal and pathological behavior of proteins.

WATER SHIELDING—SOLVENT-INDUCED INDIRECT FORCES

Water shielding of the protein surface is a crucial aspect of protein solvation, significantly impacting both structural stability and functional dynamics. The outer layer of proteins, composed mainly of polar and charged amino acids but also of hydrophobic side chains, interacts extensively with surrounding water molecules, forming a hydration shell. The water molecules in this layer form hydrogen bonds, electrostatic interactions, and van der Waals interactions with the surface residues, effectively "shielding" the protein from direct exposure to the solvent (Ball 2008, Persson et al. 2018). The interaction of the hydration shell with the side chains at the protein surface contributes to protein stabilization by water, assuming the role of bridging spatially adjacent residues. Water can bridge polar side chains through hydrogen bonds. They can also bridge hydrophobic residues through hydrogen bonds among water molecules and van der Waals interactions between water and aliphatic chains.

Water bridging at the protein surface can act as a medium for long-range interactions, facilitating communication between distant sites within

Figure 3.1 *Contd.....*
undergoing conformational exchange are colored orange or blue. Orange indicates clusters with more significant structural changes between the ground and excited states. On the right are surface hydrophobic clusters in Psd2, highlighting residues involved in conformational exchange. Residues in red are undergoing conformational exchange, whereas those in blue are not. Prolines and unassigned residues, shown in black, exhibit extreme line broadening due to conformational exchange. (C) Energy landscape of Psd2 as a function of temperature. The plot considers a three-state equilibrium: A represents the ground state, B represents the hot state, and C represents the cold destabilized state. The plot was generated from the thermodynamic parameters obtained from CPMG relaxation dispersion experiments (Pinheiro-Aguiar et al. 2023). Energy landscape temperatures: –5°C (black), 0°C (red), 5°C (blue), 10°C (green), 15°C (purple), 20°C (orange), and 25°C (cyan). Note that at low temperatures (–5°C), the C-state becomes prevalent.

the protein structure (Guan et al. 2022). This effect is particularly evident in allosteric proteins, where water molecules can transfer perturbations from one region to another, inducing conformational shifts that regulate activity at a distant site. By shielding and organizing the protein surface, water not only preserves the structural integrity of the protein but also plays an active role in its functional mechanisms, including signal transduction, molecular recognition, and enzyme activity.

The shielding effect of water on the protein surface is not static but highly dynamic, as water molecules constantly exchange with one another in the hydration shell. Magnetic relaxation dispersion experiments using (Sunde and Halle 2009) water labeled with ^{17}O nuclei have shown that most of the molecules in the hydration shell are only twofold more rigid than those in bulk water. This high mobility of the hydration shell is important to ensure rapid and efficient molecular recognition processes at the protein-water interface (Halle et al. 2004). The displacement of water molecules from the binding site can contribute to the energetics of binding. In enzyme catalysis, for example, water molecules are often involved in the active site, either participating in the reaction or stabilizing transition states. The dynamics of solvation can shift based on changes in temperature, ionic strength, or the presence of cosolvents, impacting protein activity (Ball 2008). Desolvation of the unfolded chain can also guide the local interactions that drive protein folding (Levy and Onuchic 2004). The effect of water on protein folding can be studied computationally via molecular dynamics simulations on the millisecond timescale for fast-folding proteins (Lindorff-Larsen et al. 2011).

This dynamic behavior helps accommodate the subtle conformational changes that proteins undergo during folding, binding, or catalysis. By buffering these structural changes, water ensures that the protein maintains a favorable balance between stability and flexibility. In proteins with exposed hydrophobic regions, shielding by water is particularly important because it prevents aggregation, a process that can lead to loss of function or misfolding.

The shielding enabled by the hydration shell also regulates the "dryness" of the protein core, regulating water access to internal interactions. As a result, the solvation shell indirectly induces forces at the protein core (Ben-Naim 2014, Durell and Ben-Naim 2017). Hydrogen bonds and electrostatic interactions are exacerbated if water access is restricted. In an important contribution to the stability of double-stranded DNA, Feng and colleagues reported that the primary stabilizing force of the DNA double helix is not the hydrogen bonds between base pairs but rather the coin-pile stacking of the bases. This stacking relies on hydrophobic interactions that, paradoxically, depend on the presence of ample water to create a dry environment inside the DNA, allowing hydrogen bonds to function optimally for molecular recognition. The presence of chemicals disrupting the solvation shell of the DNA double helix results in disruptions in base-pair stacking (Feng et al. 2019).

Solvent-induced forces are also present on protein surfaces. Durell and Ben-Naim calculated the interaction force between two hydrophobic molecules in an aqueous environment and reported that the main attractive force is mediated by water. They computed the potential of the mean force between 2 alkanes (methane, ethane, propane, and isobutane) as they approach each other from infinite separation to direct contact in an aqueous environment, indicating that the primary attractive forces arise from solvent-mediated interactions rather than from direct interactions between the alkanes themselves (Durell and Ben-Naim 2017).

CORELESS PROTEINS—IDEAL MODELS FOR STUDYING SURFACE HYDROPHOBIC CLUSTERS

Overall, the protein hydration shell plays a crucial role in stabilizing the protein by forming a protective barrier that regulates the "dryness" of the protein core. This shell is highly dynamic, facilitating rapid and efficient molecular recognition at the protein surface (Halle et al. 2004, Ball 2008). While most interactions in the hydration layer involve hydrophilic residues, interactions with hydrophobic residues on the protein surface are also likely to be significant in molecular recognition and allosteric regulation. These hydrophobic residues are shielded from direct water contact through interactions with neighboring residues and the surrounding hydration layer.

In recent contributions, our group highlighted that plant defensins, disintegrins, and other proteins that are highly crosslinked by disulfide bonds lack a canonical hydrophobic core, displaying all their hydrophobic residues on the protein surface, yet remain highly soluble in aqueous environments. This suggests that these residues are protected by spatially adjacent residues and the solvation layer, forming a surface hydrophobic cluster (Machado et al. 2018, Pinheiro-Aguiar et al. 2020, Vasconcelos et al. 2022, 2024). Owing to the absence of a traditional protein core, the folding of these proteins relies heavily on surface forces, making these coreless proteins ideal models for studying folding mechanisms.

Hydrophobic clusters in the protein core play critical roles in determining protein folding, stability, and interactions with other biomolecules. However, assessing the contribution of surface hydrophobic clusters to protein stability is more challenging in globular proteins, as separating their effects from other global forces, such as the forces that drive hydrophobic collapse, is difficult.

We used the plant defensin *Pisum sativum* defensin 2 (Psd2), which functions as a local and independent foldon, as a model to investigate the formation and stabilization of surface hydrophobic clusters (Pinheiro-Aguiar et al. 2023). Plant defensins contain 4 disulfide bonds and folds as typical cysteine-stabilized αβ-fold (CSαβ-fold) (Valente et al. 2013, Kovaleva et al. 2020). Despite having 4 disulfide bonds, plant defensins are generally highly dynamic, exhibiting micro- to millisecond dynamics throughout the protein, including secondary structural elements (Medeiros et al. 2010, de Paula et al. 2011, Valente et al. 2013, De Paula et al. 2013, 2014, Omidvar et al. 2016, Khairutdinov et al. 2017, Machado et al. 2018, Bukhteeva et al. 2021, 2022).

The complex dynamics of proteins cross-linked by disulfide bonds are often attributed to disulfide isomerization (Otting et al. 1993, Grey et al. 2003). We propose that this isomerization reflects a conformational equilibrium initiated at the protein surface, primarily involving surface-exposed hydrophobic residues. By computing mutational frustration via the Frustratometer server (http://frustratometer.qb.fcen.uba.ar/) (Jenik et al. 2012, Parra et al. 2016) for Psd2 (Pinheiro–Aguiar et al. 2020) and disintegrin jarastatin (Vasconcelos et al. 2024), we found that most of the frustrated contacts were concentrated at the protein surface (Figure 3.1A). Both proteins exhibit complex dynamics across multiple timescales. According to our hypothesis, these dynamics stem from the search for minimally frustrated states, which are predominantly located on the protein surface.

The complex dynamics of Psd2 and its temperature dependence enable an understanding of the thermodynamics of surface hydrophobic clusters. All of the hydrophobic residues of Psd2 (L6, F10, I15, F39, L30, and L31) face the protein surface and are involved in surface hydrophobic clusters (Figure 3.1B). Interestingly, all the clusters show residues in conformational exchange, either in the hydrophobic residue *per se* or in the spatially adjacent polar residues (Figure 3.1B).

The temperature dependence of the ^{15}N CPMG relaxation dispersion profiles reveals a correlation between residues involved in conformational exchange and surface hydrophobic clusters. Chemical shift changes between the native ground state and the first thermally accessible excited state allowed us to map the key conformational changes in Psd2 conformational equilibrium. Psd2 undergoes a three-state conformational equilibrium: the A state is the ground native state, the B state is more prevalent at higher temperatures (heat-induced state), and the C state is a cold-induced conformational state (Figure 3.1C).

The cold-driven C-state is stabilized by enthalpy and exhibits unfavorable entropy changes ($\Delta H < 0$ and $\Delta S < 0$), indicating that it exposes hydrophobic surfaces to the solvent, which is characteristic of cold denaturation (Privalov 1990). This exposure leads to a decrease in both the entropy and enthalpy. The favorable decrease in enthalpy compensates for the unfavorable reduction in entropy (Tsai et al. 2002). This enthalpy–entropy compensation is observed across both low- and high-temperature ranges.

In contrast, the hot-driven B-state shows the opposite behavior, with increases in both enthalpy and entropy ($\Delta H > 0$ and $\Delta S > 0$), similar to what has been observed for other defensins, such as Sugarcane Defensin 5 (SD5, (Machado et al. 2018)). For SD5, the excited state exhibited a significant enthalpy penalty and minimal entropic contribution. For Psd2, we observed an entropy-enthalpy compensation at 25°C.

While the C-state is primarily stabilized by enthalpy, the B-state is mainly stabilized by entropy. The C-state exposes hydrophobic surfaces, whereas the B-state protects these surfaces from solvent exposure, potentially leading to a more compact structure, as observed for SD5 (Machado et al. 2018). The balance of these effects accounts for the temperature-dependent energy landscape shown in Figure 3.1C.

In conclusion, to our knowledge, there has not been a comprehensive experimental study directly linking surface hydrophobic clusters, hydrophobic solvation, and conformational exchange. While numerous theoretical studies have explored solvent-induced surface forces (Ben-Naim 2014, Baldwin and Rose 2016, Durell and Ben-Naim 2017), hydrophobic forces remain challenging to investigate experimentally. We believe that plant defensins, disintegrins and other coreless proteins offer valuable models to advance experimental understanding in this area.

ACKNOWLEDGMENTS

We thank the Fundação Carlos Chagas Filho de Amparo à Pesquisa do Estado do Rio de Janeiro (FAPERJ) (#273303), Conselho Nacional Científico e Tecnológico (CNPq) (313517/2021-5) for funding. AAV is a postdoctoral fellow of FAPERJ "NOTA 10" and RPA is a postdoctoral fellow of FAPERJ.

REFERENCES

Almeida, F.C.L., K. Sanches, R. Pinheiro-Aguiar, V.S. Almeida and I.P. Caruso. 2021. Protein surface interactions Theoretical and experimental studies. Front. Mol. Biosci. 8: 706002.

Anfinsen, C.B., E. Haber, M. Sela and F.H. White. 1961. The kinetics of formation of native ribonuclease during oxidation of the reduced polypeptide chain. Proc. Natl. Acad. Sci. 47: 1309–1314.

Anfinsen, C.B. 1972. The formation and stabilization of protein structure. Biochem. J. 128: 737–749.

Anfinsen, C.B. 1973. Principles that govern the folding of protein chains. Science. 181: 223–230.

Arolas, J.L., S. Bronsoms, F.X. Aviles, S. Ventura and C.P. Sommerhoff. 2008. Oxidative folding of leech-derived tryptase inhibitor via native disulfide-bonded intermediates. Antioxid. Redox Signal. 10: 77–85.

Baldwin, R.L. 2014. Dynamic hydration shell restores Kauzmann's 1959 explanation of how the hydrophobic factor drives protein folding. Proc. Natl. Acad. Sci. 111: 13052–13056.

Baldwin, R.L. and G.D. Rose. 2016. How the hydrophobic factor drives protein folding. Proc. Natl. Acad. Sci. 113: 12462–12466.

Ball, P. 2008. Water as an active constituent in cell biology. Chem. Rev. 108: 74–108.

Ben-Naim, A. 2003. Hydrophobic hydrophilic phenomena in biochemical processes. Biophys. Chem. 105: 183–193.

Ben-Naim, A. 2012. Levinthal's question revisited, and answered. J. Biomol. Struct. Dyn. 30: 113–124.

Ben-Naim, A. 2014. Water's contribution in providing strong solvent-induced forces in protein folding. Eur. Phys. J. Spec. Top. 223: 927–946.

Berardi, M.J., W.M. Shih, S.C. Harrison and J.J. Chou. 2011. Mitochondrial uncoupling protein 2 structure determined by NMR molecular fragment searching. Nature. 476: 109–113.

Bhopatkar, A.A., V.N. Uversky and V. Rangachari. 2020a. Disorder and cysteines in proteins: A design for orchestration of conformational see-saw and modulatory functions. Prog. Mol. Biol. Transl. Sci. 174: 331–373.

Bhopatkar, A.A., V.N. Uversky and V. Rangachari. 2020b Granulins modulate liquid–liquid phase separation and aggregation of the prion-like C-terminal domain of the neurodegeneration-associated protein TDP-43. J. Biol. Chem. 295: 2506–2519.

Bhopatkar, A.A., S. Dhakal, H.G. Abernathy, S.E. Morgan and V. Rangachari. 2022. Charge and redox states modulate granulin—TDP-43 coacervation toward phase separation or aggregation. Biophys. J. 121: 2107–2126.

Bryngelson, J.D. and P.G. Wolynes. 1987. Spin glasses and the statistical mechanics of protein folding. Proc. Natl. Acad. Sci. 84: 7524–7528.

Bryngelson, J.D., J.N. Onuchic, N.D. Socci and P.G. Wolynes. 1995. Funnels, pathways, and the energy landscape of protein folding: A synthesis. Proteins Struct. Funct. Bioinforma. 21: 167–195.

Bukhteeva, I., N.I. Hrunyk, Y.M. Yusypovych, Y.I. Shalovylo, V. Kovaleva and I.V. Nesmelova. 2022. Structure, dynamics, and function of PsDef2 defensin from *Pinus sylvestris*. Structure. 30: 753–762.

Craik, D.J. 2011. The folding of disulfide-rich proteins. Antioxid. Redox Signal. 14: 61–64.

Cruzeiro-Silva, C., F. Gomes-Neto, L.E. Machado, C.A. Miyamoto, A.S. Pinheiro, N. Correa-Pereira, et al. 2014. Hydration and conformational equilibrium in yeast thioredoxin 1: implication for H^+ exchange. Biochemistry. 53: 2890–2902.

De Paula V.S., G. Razzera, E. Barreto-Bergter, F.C.L. Almeida and A.P. Valente. 2011. Portrayal of complex dynamic properties of sugarcane defensin 5 by NMR: multiple motions associated with membrane interaction. Structure. 19: 26–36.

De Paula, V.S., N.S. Gomes, L.G. Lima, C.A. Miyamoto, R.Q. Monteiro, F.C. Almeida, et al. 2013. Structural basis for the interaction of human β-defensin 6 and its putative chemokine receptor CCR2 and Breast cancer microvesicles. J. Mol. Biol. 425(22): 4479–4495.

De Paula V.S., V.H. Pomin and A.P. Valente. 2014. Unique properties of human β-defensin 6 (hBD6) and glycosaminoglycan complex: Sandwich-like dimerization and competition with the chemokine receptor 2 (CCR2) binding site. J. Biol. Chem. 289: 22969–22979.

Delaunay-Moisan, A., A. Ponsero and M.B. Toledano. 2017. Reexamining the function of glutathione in oxidative protein folding and secretion. Antioxid. Redox Signal. 27: 1178–1199.

Dill, K.A. and J.L. MacCallum. 2012. The protein-folding problem, 50 years on. Science. 338: 1042–1046.

Durell, S.R. and A. Ben-Naim. 2017. Hydrophobic-hydrophilic forces in protein folding. Biopolymers. 107: e23020.

Erdos, G. and Z. Dosztányi. 2024. AIUPred: Combining energy estimation with deep learning for the enhanced prediction of protein disorder. Nucleic. Acids. Res. 52: W176–W181.

Esperante, S.A., G. Covaleda, S. A. Trejo, S. Bronsoms, F.X. Aviles and S. Ventura. 2017. Plasticity in the oxidative folding pathway of the high affinity nerita versicolor carboxypeptidase inhibitor (NvCI). Sci. Rep. 7: 1–16.

Feng, B., R.P. Sosa, A.K.F. Mårtensson, K. Jiang, A. Tong, K.D. Dorfman, et al. 2019. Hydrophobic catalysis and a potential biological role of DNA unstacking induced by environment effects. Proc. Natl. Acad. Sci. 116 (35): 17169–17174.

Ferreiro, D.U., E. Komives and P.G. Wolynes. 2014. Frustration in biomolecules. Q. Rev. Biophys. 47(4): 285–363.

Fraga, H. and S. Ventura. 2012. Protein oxidative folding in the intermembrane mitochondrial space: More than protein trafficking. Curr. Protein. Pept. Sci. 13: 224–231.

Ghag, G., L.M. Wolf, R.G. Reed, N.P. Van Der Munnik, C. Mundoma, M.A. Moss, et al. 2016. Fully reduced granulin-B is intrinsically disordered and displays concentration-dependent dynamics. Protein. Eng. Des. Sel. 29: 177–186.

Ghag, G., C.J. Holler, G. Taylor, T.L. Kukar, V.N. Uversky and V. Rangachari, V. 2017. Disulfide bonds and disorder in granulin-3: An unusual handshake between structural stability and plasticity. Protein Sci. 26: 1759–1772.

Gianni, S., M.I. Freiberger, P. Jemth, D.U. Ferreiro, P.G. Wolynes and M. Fuxreiter. 2021. Fuzziness and frustration in the energy landscape of protein folding, function, and assembly. Acc. Chem. Res. 54: 1251–1259.

Grey, M.J., C. Wang and A.G. Palmer. 2003. Disulfide bond isomerization in basic pancreatic trypsin inhibitor: Multisite chemical exchange quantified by CPMG relaxation dispersion and chemical shift modeling. J. Am. Chem. Soc. 125: 14324–14335.

Guan, X., C. Tan, W. Li, W. Wang and D. Thirumalai. 2022. Role of water-bridged interactions in metal ion coupled protein allostery. PLoS Comput. Biol. 18: 1–20.

Halle, B., J.R. Helliwell, A. Kornyshev and J.B.F.N. Engberts. 2004. Protein hydration dynamics in solution: A critical survey. Philos. Trans. R. Soc. B. Biol. Sci. 359: 1207–1224.

Iqbal. A., F. Gomes-Neto, C.A. Myiamoto, A.P. Valente and F.C.L. Almeida. 2015. Dissection of the water cavity of yeast thioredoxin 1: The effect of a hydrophobic residue in the cavity. Biochemistry 54: 2429–2442.

Janin, J. 1979. Surface and inside volumes in globular proteins. Nature. 277: 491–492.

Jenik, M., R.G. Parra, L.G. Radusky, A. Turjanski, P.G. Wolynes and D.U. Ferreiro. 2012. Protein frustratometer: A tool to localize energetic frustration in protein molecules. Nucleic Acids Res. 40: 348–351.

Jorgensen, W.L., J. Gao and C. Ravimohan. 1985. Monte Carlo simulations of alkanes in water: Hydration numbers and the hydrophobic effect. J. Phys. Chem. 89: 3470–3473.

Kadokura, H., F. Katzen and J. Beckwith. 2003. Protein disulfide bond formation in prokaryotes. Annu. Rev. Biochem. 72: 111–135.

Kauzmann, W. 1959. Some factors in the interpretation of protein denaturation. Adv. Protein. Chem. 14: 1–63.

Khairutdinov, B.I., E.A. Ermakova, Y.M. Yusypovych, E.K Bessolicina, N.B. Tarasova, Y.Y Toporkova, et al. 2017. NMR structure, conformational dynamics, and biological activity of PsDef1 defensin from Pinus sylvestris. Biochim. Biophys. Acta. Proteins Proteomics. 1865: 1085–1094.

Kovaleva, V., I. Bukhteeva, O.Y. Kit and I.V. Nesmelova. 2020. Plant defensins from a structural perspective. Int. J. Mol. Sci. 21: 5307.

Kulkarni, P., S. Bhattacharya, S. Achuthan, A. Behal, M.K. Jolly, S. Kotnala, et al. 2022. Intrinsically disordered proteins: Critical components of the wetware. Chem. Rev. 122: 6614–6633.

Leopold, P.E., M. Montal and J.N. Onuchic. 1992. Protein folding funnels: A kinetic approach to the sequence-structure relationship. Proc. Natl. Acad. Sci. 89: 8721–8725.

Levinthal, C. 1968. Are there pathways for protein folding? J. Chim. Phys. 65: 44–45.

Levy, Y. and J.N. Onuchic. 2004. Water and proteins: A love-hate relationship. Proc. Natl. Acad. Sci. 101: 3325–3326.

Lindorff-Larsen, K., S. Piana, R.O. Dror and D.E. Shaw. 2011. How fast-folding proteins fold. Science. 334: 517–520.

Long, D., G. Bouvignies and L.E. Kay. 2014. Measuring hydrogen exchange rates in invisible protein excited states. Proc. Natl. Acad. Sci. 111: 8820–8825.

Machado, L.E.S.F., V.S. De Paula, Y. Pustovalova, I. Bezsonova, A.P.Valente, D.M. Korzhnev, et al. 2018. Conformational Dynamics of a cysteine-stabilized plant defensin reveals an evolutionary mechanism to expose hydrophobic residues. Biochemistry. 57(40): 5797–5806.

Malhotra, P. and J.B. Udgaonkar. 2016. How cooperative are protein folding and unfolding transitions? Protein. Sci. 25: 1924–1941.

Medeiros, L.N., R. Angeli, C.G. Sarzedas, E. Barreto-Bergter, A.P. Valente, E. Kurtenbach, et al. 2010. Backbone dynamics of the antifungal Psd1 pea defensin and its correlation with membrane interaction by NMR spectroscopy. Biochim. Biophys. Acta. 1798: 105–113.

Mészáros, B., G. Erdös and Z. Dosztányi. 2018. IUPred2A: Context-dependent prediction of protein disorder as a function of redox state and protein binding. Nucleic Acids Res. 46: W329–W337.

Modig, K., E. Liepinsh, G. Otting and B. Halle. 2004. Dynamics of protein and peptide hydration. J. Am. Chem. Soc. 126: 102–114.

Nucci, N.V., M.S. Pometun and A.J. Wand. 2011 Mapping the hydration dynamics of ubiquitin. J. Am. Chem. Soc. 133: 12326–12329.

Omidvar, R., Y. Xia, F. Porcelli, H. Bohlmann and G. Veglia. 2016. NMR structure and conformational dynamics of AtPDFL2.1, a defensin-like peptide from Arabidopsis thaliana. Biochim. Biophys. Acta, Proteins Proteomics. 1864: 1739–1747.

Onuchic, J.N., Z. Luthey-Schulten and P.G.Wolynes. 1997. Theory of protein folding: The energy landscape perspective. Annu. Rev. Phys. Chem. 48: 545–600.

Otting, G., E. Liepinsh and K. Wüthrich. 1991. Protein hydration in aqueous solution. Science. 254: 974–980.

Otting, G., E. Liepinsh and K. Wüthrich. 1993. Disulfide bond isomerization in BPTI and BPTI (G36S): An NMR study of correlated mobility in proteins. Biochemistry. 32: 3571–3582.

Otting, G., E. Liepinsh, B. Halle and U. Frey. 1997. NMR identification of hydrophobic cavities with ow water occupancies in protein structures using small gas molecules. Nat. Struct. Mol. Biol. 4: 396–404.

Palmer, A.G. 2014. Chemical exchange in biomacromolecules: past, present, and future. J. Magn. Reson. 241: 3–17.

Parra, R.G., N.P. Schafer, L.G. Radusky, M.Y. Tsai, A.B. Guzovsky, P.G. Wolynes, et al. 2016. Protein Frustratometer 2: a tool to localize energetic frustration in protein molecules, now with electrostatics. Nucleic Acids Res. 44: W356–W360.

Persson, F., P. Söderhjelm and B. Halle. 2018. The spatial range of protein hydration. J. Chem. Phys. 148: 215104.

Pinheiro-Aguiar, R., V.S.G. do Amaral, I.B. Pereira, E. Kurtenbach and F.C.L. Almeida. 2020. Nuclear magnetic resonance solution structure of *Pisum sativum* defensin 2 provides evidence for the presence of hydrophobic surface-clusters. Proteins Struct. Funct. Bioinforma. 88: 242–246.

Pinheiro-Aguiar, R., C. Corrêa-Almeida, E. Kurtenbach and F.C.L. Almeida. 2023. Plant defensins as a model to study the thermostability of surface hydrophobic clusters: The *Pisum sativum* defensin 2 (Psd2). J. Magn. Reson. Open 16–17: 100106.

Privalov, P.L. 1990. Cold denaturation of proteins. Biocemistry Annu. Rev. 25: 281–305.

Qvist, J. and B. Halle. 2008. Thermal signature of hydrophobic hydration dynamics. J. Am. Chem. Soc. 130: 10345–10353.

Stelzl, L.S., D.A. Mavridou, E. Saridakis, D. Gonzalez, A.J. Baldwin, S.J. Ferguson, et al. 2020. Local frustration determines loop opening during the catalytic cycle of an oxidoreductase. Elife. 9: 1–27.

Sunde, E.P. and B. Halle. 2009. Slow internal protein dynamics from water 1 H magnetic relaxation dispersion. J. Am. Chem. Soc. 131: 18214–18215.

Tsai, C.J., J.V. Maizel and R. Nussinov. 2002. The hydrophobic effect: A new insight from cold denaturation and a two-state water structure. Crit. Rev. Biochem. Mol. Biol. 37: 55–69.

Valente, A.P., V. de Paula and F.C.L. Almeida. 2013. Revealing the properties of plant defensins through dynamics. Molecules. 18: 11311–11326.

Vasconcelos, A.A., J.C. Estrada, V. David, L.S. Wemelinger, F.C.L. Almeida and R.B. Zingali. 2021a. Structure-function relationship of the disintegrin family: sequence signature and integrin interaction. Front. Mol. Biosci. 8: 1–20.

Vasconcelos, A.A., B.B. Succar, L.B. di Piero, E. Kurtenbach, R.B. Zingali, F.C.L. Almeida. 2022. 15N, 13C, and 1H resonance assignments of Jarastatin: A disintegrin of Bothrops jararaca. Biomol. NMR Assign. 16: 37–40.

Vasconcelos, A.A., J.C. Estrada, I.P. Caruso, E. Kurtenbach, R.B. Zingali and F.C.L. Almeida. 2024. Toward the mechanism of jarastatin (rJast) inhibition of the integrin $\alpha V\beta 3$. Int. J. Biol. Macromol. 255: 128078.

Vinci, F., S. Catharino, S. Frey, J. Buchner, G. Marino, P. Pucci, et al. 2004. Hierarchical formation of disulfide bonds in the immuno-globulin Fc fragment is assisted by protein-disulfide isomerase. J. Biol. Chem. 279: 15059–15066.

Wang, L. and Wang, C. 2023. Oxidative protein folding fidelity and redoxtasis in the endoplasmic reticulum. Trends Biochem. Sci. 48: 40–52.

Whittaker, S.B M., N.J. Clayden and G.R. Moore. 2011. NMR characterisation of the relationship between frustration and the excited state of Im7. J. Mol. Biol. 414: 511–529.

Wolfenden, R., C.A. Lewis, Y. Yuan and C.W. Carter. 2015. Temperature dependence of amino acid hydrophobicities. Proc. Natl. Acad. Sci. 112: 7484–7488.

Wolynes, P.G. 2004. Energy landscapes and solved protein-folding problems. Philos. Trans. R. Soc. A. Math. Phys. Eng. Sci. 363: 453–467.

Wolynes, P.G. 2015. Evolution, energy landscapes and the paradoxes of protein folding. Biochimie. 119: 218–230.

Zarges, C. and J. Riemer. 2024. Oxidative protein folding in the intermembrane space of human mitochondria. FEBS Open Bio. 14: 1610–1626.

Zhou, H. and Y. Zhou. 2004. Quantifying the effect of burial of amino acid residues on protein stability. Proteins. 54: 315–322.

Chapter **4**

Residue Cluster Classes in Protein-Protein Interactions

Juan Martín Hernández Castillo[1], Albros Hermes Poot[1] and Gabriel Del Río*[1]

[1]Department of Biochemistry and Structural Biology, Instituto de Fisiologia Celular, UNAM, Mexico City 04510, Mexico

INTRODUCTION

Protein-Protein Interaction (PPI) prediction constitute a valuable tool to assist researchers to identify a potential molecular complex associated with a particular molecular process (e.g., metabolism, signal transduction, among others). Given the relevance of such tasks and the relevance of understanding how such interactions evolved, there is an ongoing international effort to evaluate methods that predict PPI using protein three-dimensional (3D) structure (Lensink et al. 2019).

Besides the practical application of PPI predictions, these predictors may be used to evaluate the relevance of current models used to explain the PPI phenomena. For instance, physicochemical factors (e.g., charge, hydrophobicity) are commonly used to explain PPI, but are these factors the most effective to predict PPI? In the past two decades, several discoveries about protein structure and function are promoting the development of new paradigms/models to understanding the

* For Correspondence: gdelrio@ifc.unam.mx

protein structure-function relationship. Examples of such paradigm-shifting discoveries are AlphaFold (Tunyasuvunakool et al. 2021) and Intrinsically Disordered Regions of proteins (IDRs) or Intrinsically Disordered Proteins (IDPs) (Dunker et al. 2001, Malagrinò et al. 2022, Uversky et al. 2000, Wright and Dyson 1999). AlphaFold uses a two-step process to infer the final 3D structure of the protein (Jumper et al. 2021). In the first step, AlphaFold generates a matrix describing the potential contact map of amino acids for the protein of interest and, in the second step, AlphaFold finds the backbone that satisfies such a contact map. AlphaFold differs from the idea that physicists worked on for many years, where the final protein structure would be identified by sampling the available conformational space of proteins using different force-fields. This physics-based idea attempts to solve the protein folding problem, while AlphaFold solves the problem of identifying the preferred final structure of the protein.

In this regard, AlphaFold can predict a single stable conformation adopted for some proteins, but also to predict IDRs in proteins (Alderson et al. 2023). However, AlphaFold has limited information on metamorphic proteins (proteins that switch structures in physiological conditions) and at the moment cannot be used to predict such proteins (Chakravarty et al. 2024, Porter et al. 2024). IDRs or IDPs do not present a unique secondary or tertiary structure, hence constitute regions/domains that may adopt multiple conformations. This has led to propose a shift in the previous paradigm of protein folding, where a single global minimum of energy was assumed and now, multiple equally likely global minimum may occur during protein folding that may be shifted by a ligand (Malagrinò et al. 2022). Metamorphic proteins are also examples of proteins where a roughed protein folding landscape allows the occurrence of more than one stable conformation (Dishman and Volkman 2022). Thus, new concepts/frameworks are required to understand IDRs, metamorphic proteins, IDPs and structured proteins to predict how many minima of energy may occur in every protein. The continuum model proposes that proteins may adopt from a single to multiple minima of energy in a continuum fashion (Gupta and Uversky 2024). The continuum model represents an important new perspective to understand protein folding and function, and it may contribute to predict when a given protein sequence may have one or many minima of energy and functions.

In the present work, we will analyze an emerging idea related to clusters of amino acids in proteins, where such clusters are considered the building blocks of protein domains and interacting proteins. Under such a framework, these building blocks will be formed depending on the protein sequence composition, leading to one or multiple ways such packing blocks adopt one or multiple protein structures or PPI. In the first section, the residue clusters referred in this study will be described. Also included is the description of how Residue Cluster Classes (RCC) have been used to represent the 3D structure of proteins in a vector and the efficacy of this approach to classify protein structure, function, and predict PPI. In the second section we will compare the efficiency of PPI prediction by RCC with other methods to evaluate the significance of using such a cluster of residues to describe PPI. Finally, in the third section we will describe the orthogonality of those proteins forming a PPI and those that do not form a PPI, to explain how a cluster of residues in proteins may be organized to facilitate PPI. Based on these analyses, we propose RCC as an efficient and simple representation of protein structure to model the structure-function relationship of proteins.

RESIDUE CLUSTER CLASSES IN PROTEINS

A protein sequence *P*, with a set of residues *r*, that upon folding may adopt a conformation *C*. From such a conformation, every pair of residues that are in contact at a given distance *d* define the set of edges at a given distance, e_d, for that conformation of the given protein. Thus, $r(P_i)$ refers to the residues found in *i*-protein; $r(C_i)$ refers to the residues present in the contact map at a given distance defining the protein's *i*-conformation; $r1_i$, $r2_i$, $r3_i$, ..., rn_i refers to the list of residues in the *i*-protein or *i*-conformation; $e(C_i)$ refers to the edges found at the protein's *i*-conformation; (ra_i, rb_i) refers to a pair of residues making up an edge as defined in the protein's *i*-conformation.

$$r(P_i) = \{r1_i, r2_i, r3_i, \ldots, rn_i\},\ r(C_j) = \{r1_j, r2_j, r3_j, \ldots, rn_j\}$$

while

$$e(Cj) = \{(ra_j, rb_j), (rc_j, rd_j), \ldots, (rx_j, ry_j)\}$$

$$e(C_j) \subset r(C_j)$$

$$P = \{[r(C_i), e(C_j)], [r(C_m), e(C_n)], \ldots, [r(C_x), e(C_y)]\}$$

Please note that for a given protein sequence *P*, there may be more than one conformation *C*, hence different sets of edges may be used to described different protein conformations.

Given the sets of residues *r* and edges *e* for a conformation *C*, Residue Clusters (RC) are identified as maximal cliques, or cliques that cannot be extended by including an adjacent residue, that is a maximal clique is not a subset of another clique. Thus, RC on *P* is a subset $A \subseteq 2^R$, such that $A \subseteq e(C)$; A corresponds to the set of non-hydrogen atoms of the residues. The Residue Cluster Class (RCC) system corresponds to a Sperner family (Lubell 2009), which is defined as a set of all residue clusters for all k-Maximal Cliques, k-MC, such that none is subset of some other. The k refers to the size of the maximal cliques, which is the number of residues in the clique. We defined different classes of maximal cliques based on the presence of residues that are adjacent on the protein sequence (contiguous at the protein sequence); for instance, a maximal clique of size 3 could have (i) 3 adjacent residues (refer to as [3]), and (ii) 2 adjacent residues (refer to as [1,2]), (iii) 3 non-adjacent residues (refer to as [1,1,1]). For maximal cliques of size 3, 4, 5 and 6 we identified 26 different maximal cliques based on the presence of contiguous residues (Corral-Corral et al. 2015). Thus, a vector of RCC (vRCC) resides in a 26-dimensional space, $Z+^{26}$, where Z+ corresponds to the set of positive integers, and is defined as follows:

$$\text{vRCC} = \{3\text{-MC}_1, 3\text{-MC}_2, 3\text{-MC}_3, \ldots, 6\text{-MC}_{26}\}$$

Where 3-MC_1 refers to the first class of 3-MC, 3-MC_2 to the second class of 3-MC and so on. Such vRCC accounts for the densest regions of proteins.

Please note that the ordered collections of RCC do not describe a vector space, hence the term vector is used here to facilitate the use of operations commonly associated to vectors.

We have reported that vRCC is an effective representation of protein structure, such that protein structure classification based on CATH or SCOP criteria are reproduced with better reliability than any other method (Corral-Corral et al. 2015). vRCC matched or improved the reliability of known protein function prediction methods reported in CAFA (Fontove and Del Rio 2020) and vRCC was effective in identifying pairs of interacting from non-interacting proteins (Poot Velez et al. 2020). While in these previous studies on protein structure and protein function classification we have made a comparison with other methods, we have not compared the efficiency of our method to predict PPI with current state-of-the-art PPI predictors, which would be relevant to propose a framework to understand the role RCC plays in understanding the protein structure and function.

Here we will test if RCC fit the Occam's razor criteria to represent proteins and model the protein structure and function, that is we will analyze if RCC reduce the complexity of models for protein structure and function while improving or maintaining the prediction quality. PPI include both, protein structure and function, thus comparing RCC to predict PPI with other models represents an appropriate way to test the Occam's razor criteria. In the next section, we provide a theoretical framework to help interpreting the reasoning behind RCC to model protein structure and function.

RCC INTERPRETATION IN PROTEIN FOLDING AND PPI

The work by Ken Dill proposed for the first time a lattice model to explore the folding of proteins (Dill 1985). Such a computational model provided a framework to test for the forces at play during protein folding, particularly the solvophobic forces. While an important model, such approximation was not able to identify the final 3D structure of medium-size or large proteins (MacCallum et al. 2015). In terms of identifying the arrangement that achieves the densest packing, the Kepler conjecture on the packing problem formulated in 1611 and finally validated in 1997 by Thomas Hale (Hales 1997a, Hales 1997b), proposed that the Face-Centered-Cubic (FCC) lattice renders the densest arrangement for identical spheres in 3D. FCC-like arrangements have been reported in many different classes of globular protein structures, suggesting that non-identical protein side-chains follow the same arrangement as equal spheres (Bagci et al. 2002, Lara Ortiz et al. 2021). That is, protein folds minimizing the voids when packing the amino acid residues.

If proteins adopt the densest packing and RCC accounts for such dense packing, the observation that RCC correctly classify protein structures suggests that different protein structures use different sets of maximal cliques to achieve the FCC-like arrangement. This is a problem similar to irregular tessellations. A tessellation in mathematics can be defined as the filling of a surface or volume, without overlaps or gaps, by using one or more shapes (Kizilörenli and Maden 2021). Please note that while tessellation do not allow voids, proteins may as well be represented without voids and have been represented as tessellations before (Dupuis et al. 2005). Furthermore, while some regions of proteins may have voids, the RCC only accounts for those regions where it is unlikely to find voids. Additionally, an irregular tessellation combines non-regular shapes. If the non-regular shapes are infinite, the number of different irregular tessellations is infinite as well, yet if the number of irregular shapes is finite, it is not trivial to identify how many tessellations can be formed, but we can assume that these are finite as well. In that sense, the limited number of different amino acids would impose a limited number of different RCC shapes, thus the number of different irregular tessellations in proteins (that is, the number of different protein 3D structures) are finite. In such a case, the finite number of irregular tessellations observed in proteins may have patterns that are associated with the observed structural classification of proteins, as we have shown earlier (Corral-Corral et al. 2015). Thus, different sets of residues arranged in maximal cliques have different shapes, and the protein folding problem may consist in arranging such maximal cliques in a way to create a tessellation. A protein may adopt only one tessellation (one 3D structure) when there is only one way to arrange the maximal cliques formed during folding or may adopt two or more tessellations (two or more 3D structures) when multiple arrangements of the maximal cliques formed during folding render multiple tessellations. Therefore, considering protein folding as a tessellation problem, and RCC as the shapes used to form the tessellation, this provides a framework to explain the existence of IDRs and metastable proteins. This modeling has not been attempted up to date, and it is not the goal of this work. Yet, we propose this model for the first time to interpret the results that will be presented here.

If we consider that a protein single subunit should avoid voids during folding, is it possible that a similar process drives PPI? Since each protein subunit in a PPI may be considered an irregular tessellation, a protein complex also forms an irregular tessellation, that is the subunit in the complex would associate as well without allowing voids.

While there is not a solution to predict the number and forms of different irregular tessellations that can be formed for a given set of irregular shapes, we anticipate that one way to solve such a problem would be through divergence, that is to assume that once an irregular tessellation exists, a new one can be created from the one already present in nature by simply eliminating some irregular shapes and substitute those with a new set of irregular shapes that are isomorphic, that is that would fill in the missing space left by the eliminated irregular shapes (see Figure 4.1).

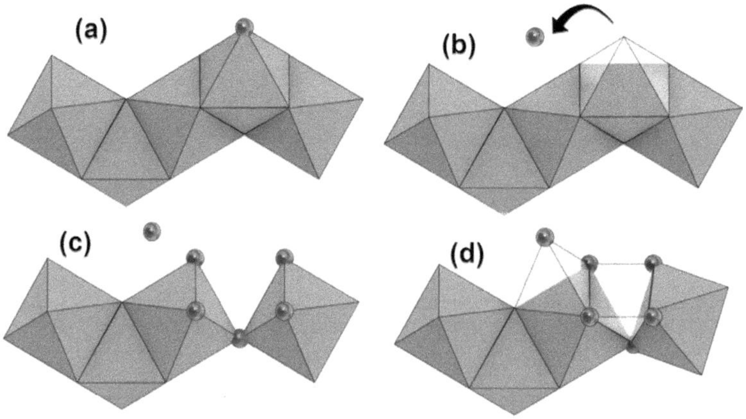

Figure 4.1 Evolution of new irregular tessellation through parallelism. A protein represented as a tessellation formed by maximal cliques. (a) Maximal cliques of size 6 (octahedrons) where in each vertex there is an amino acid. (b) One amino acid changes its position, breaking the clique where it originally belonged. (c) The protein search to maximize the interactions that stabilize it. (d) New maximal cliques are formed, one of size 5 (oblique square pyramid) and one of size 4 (tetrahedron).

Such a process would reduce the complexity of the search to identify irregular shapes to fill in the space. If that is the case, we would expect to find a higher probability of parallelism among the RCC vectors from pairs of interacting proteins than randomly generated RCC vectors where the likelihood of having an interaction for a protein is equally distributed. Two RCC vectors are parallel if they share some types of RCC (same positions in the RCC vector different from zero) and differ in some others; hence, such proteins should have preserved some packing blocks (maximal cliques) as exemplified in Figure 4.1. We would not expect to find that all pairs of RCC in the current sample of known PPI to be parallel, because we anticipate other ways to generate new PPI not based on creating parallelism; such would be cases of convergence that may not generate parallelism between the RCC from each protein. Also, not all possible conformations for a protein (nor all different protein 3D structures) are included in our analysis, hence some parallels will be missed. Please note that among PPI are pairs of identical proteins, where our assumption is satisfied.

In the next section, we will explore the efficiency of different methods to predict PPI and the data sources used for such predictions. We will compare those efficiencies with that reported for PPI predictions based on RCC. The goal is to test whether the Occam's razor criteria to represent PPI applies to RCC, that is, whether RCC reduces the complexity of PPI models while improving or maintaining the prediction quality.

PROTEIN-PROTEIN INTERACTION PREDICTORS

The current Critical Assessment of Structure Prediction (CASP) taking place in 2024, is evaluating PPI predictions instead of separating this problem in the CAPRI format. Hence, new information about the reliability of AlphaFold on PPI will become available by the end of this year (Bret et al. 2024, Evans et al. 2022). Since IDRs are relevant for PPI, these two emerging concepts, deep learning and IDR, may be put onto a test during CASP16.

Recent works have independently evaluated the reliability of currently available PPI predictors (Kewalramani et al. 2024, Wang et al. 2023, Xian and Wang 2024). Supplementary Tables S1, S2 and Table 4.1 summarize the datasets, number of features used, and efficiencies reported in such works, respectively.

In agreement with previous observations (Zhang et al. 2024), we found that, on average, Machine Learning (ML) methods are the most reliable classifiers of PPI (see Table 4.1). But ML methods also are those using the largest number of features to train their models (see Table S2). Considering the Occam's razor criteria, the proper model

Table 4.1 PPI prediction efficiency comparison of different models

	AUC ROC	Accuracy	Precision	Sensitivity	Specificity	Matthew's Correlation Coefficient
LightGBM-PPI	97	92	93	91	93	0.842
GcForest-PPI	98	92	93	NR	NR	0.844
FWRF	91.6	91.9	92.5	91.2	NR	0.8384
Profppikernel	74.5	NR	NR	NR	NR	NR
Goktepe et al.	89.9	83.6	81.9	81.2	NR	0.792
Hu et al.	76	70	NR	68.6	70.6	0.583
SSWRF	NR	67.9	32.2	NR	69.7	0.234
PrePPI	77	NR	0.5	NR	NR	NR
InterPred	<70	NR	NR	NR	NR	NR
Dong et al.	0.72	NR	NR	NR	NR	NR
Bryant et al.	0.9	NR	NR	0.6	NR	NR
Cluspro	NR	NR	NR	87.5	NR	NR
HADDOCK	NR	NR	NR	73.3	NR	NR
L3	NR	NR	66	NR	NR	NR
SFCN	90.9	NR	48.6	NR	NR	NR
Sim	NR	NR	35.5	NR	NR	NR
Lei et al.	NR	NR	NR	NR	NR	NR
L3N	NR	NR	NR	NR	NR	NR
PPI-MetaGO	93.8	90.3	96	NR	NR	0.785
Bandyopadhyay et al.	89.2	NR	NR	NR	NR	NR
DeepPPI	97.43	92.5	94.4	NR	94.5	0.85
Struct2Graph	99.5	98.89	99.5	NR	99.45	0.978
TAGPPI	NR	98.5	99.2	98	99.2	0.971
SGPPI	75.7	NR	31.4	44.4	NR	NR
Sun et al.	NR	97.2	96.2	98	96.3	NR
D-SCRIPT	82.6	NR	78	33.9	NR	NR
DPPI	NR	95.8	96.7	92.2	NR	NR
PPI-RCC	97	NR	96.5	83.3	95	NR

The values reported in this table correspond with the average of the best reported efficiencies. That is, if the work reported different training or testing sets, we report here the average of the best values. NR stands for Not Reported. PPI-RCC and several other predictors used the same set of PPI from 4 different organisms (*Arabidopsis thaliana, Caenorhabditis elegans, Homo sapiens* and *Saccharomyces cerevisiae*); we used that dataset (https://github.com/spxuw/PPI-Prediction-Project) to build a test set for our PPI-RCC method (https://github.com/gdelrioifc/PPI-RCC2); we referred to it as PPI-positive and as the negative set we used Negatome 2 (Blohm et al. 2014) that was not included in the training of our method, referred to as PPI-negative set. Based on the previous efficiencies reported in our work to predict PPI based on RCC, we used 4 trained algorithms (https://github.com/gdelrioifc/PPI-RCC2, with the names raw_7noSC_con_tra_sz3_it2, raw_8noSC_con_tra_sz3_it3, raw_8noSC_con_tra_sz2_it1 and raw_8noSC_sum_tra_sz2_it1) to classify the PPI-positive from PPI-negative sets. AUC ROC stands for Area Under the Curve Receiver Operating Characteristic. The methods in the table are grouped from top to bottom, as sequence-based (▒), structure-based (▒), network-based (▒), GO-based ( ) and machine-learning methods (□).

should be that with the lowest number of variables (see Table S2) and highest reliability in their predictions (see Table 4.1). Looking at Table 4.1, AUC ROC is the parameter most frequently reported by most authors and include both true and false predictions. We will use sensitivity and specificity when available to further analyzed the reliability of the prediction. Using AUC ROC, we identified Struct2Graph (99.5), followed by GcForest-PPI (98), DeepPPI (97.4), LightGBM-PPI (97) and PPI-RCC (97) as the most reliable predictors. From Table S2, we observed that network-based SFCN (12), structure-based HADDOCK (14), structure-based InterPred (17), structure-based PPI-RCC (26) and sequence-based SSWRF (50) included the lowest numbers of features to make their predictions; that is, structure-based methods use the lowest number of features to achieve PPI predictions, and PPI-RCC is the most reliable one among them.

While in this study PPI-RCC used the same dataset used by most other methods analyzed here, not all the methods used the same sets of positive and negative instances of PPI (see Table S1). Therefore, it is relevant to note the differences in datasets used to train each of these methods (see Table S1). DeepPPI includes a balanced and well tested method (using different sets of proteins in the training and testing sets), hence may be considered as the best PPI classifier with an AUC ROC value of 97.4 followed by PPI-RCC with 97. PPI-RCC on the other hand, requires much less variables (26) than DeepPPI (>500).

ORTHOGONALITY OF INTERACTING AND NON-INTERACTING PROTEINS

To identify how similar two vRCC are to each other, we determine the angle between them. The angle θ between any two vectors, $\vec{v} = (v_1, v_2, \ldots, v_n)$ and $\vec{u} = (v_1, v_2, \ldots, v_n)$ in an n-dimensional Euclidean space, can be derived using the following equation:

$$\cos(\theta) = \frac{(\vec{v} \cdot \vec{u})}{vu} = \frac{\Sigma_{i=1}^{n}(v_i u_i)}{\sqrt{\Sigma_{i=1}^{n} v_i^2}\sqrt{\Sigma_{i=1}^{n} u_i^2}}$$

Two identical proteins can be associated with an ideal case of parallelism between the vectors, $\theta = 0°$. We wonder whether interacting proteins would also share the same type of maximal cliques, hence forming an angle close to zero.

To answer this question about parallelism and PPIs, the angular distribution of vRCC was analyzed. We built contact maps at 7 and 8 Å, with and without the atoms from the side chains, considering that the best representation of PPI was achieved using such conditions (Poot Velez et al. 2020). Three cases were analyzed (data available at https://github.com/gdelrioifc/PPI-RCC2):

1. **Experimental_DB:** All proteins with experimentally determined structure and experimentally determined interaction from four organisms (Mosca et al. 2013): *Caenorhabditis elegans, Arabidopsis thaliana, Homo sapiens,* and *Saccharomyces cerevisiae.* In this set, 3320 pairs of interacting proteins (1717 unique proteins) were identified with non-null vRCCs for 7 Å with and without side chains, and 8 Å with and without side chains as previously reported (Fontove and Del Rio 2020).
2. **AlphaFold_DB:** All non-filtered proteins with structure determined by AlphaFold by May 2024 (Varadi et al. 2022) and experimentally determined interaction for the same four organisms in Experimental_DB set. In this set, the same 3307 pairs of interacting proteins (the same 1713 unique protein included in the Experimental_DB set) were identified.
3. **Random_DB:** Randomly generated vectors were created by following the average distribution of vRCCs in the reference set (Experimental_DB or AlphaFold_DB). That is, for each protein in each set, a corresponding randomly generated vRCC was created and paired according to the reference set (e.g. Experimental_DB). A total of unique 1717 random vRCC were used, to generate 3320 pairs.

For the non-interacting pairs in the Experimental_DB and AlphaFold_DB, 1717 unique proteins were sampled in groups of 83 unique proteins and paired with any other protein that was not part of the Experimental_DN and AlphaFold_DB sets. That was performed 440 times.

The angle distribution for these three sets and corresponding non-interacting pairs using 7 and 8 Å with and without side chains are shown in Figure 4.2.

Note that the angle between pairs of interacting proteins (continuous lines) with or without sidechains from the Experimental_DB shows a maximum around 25°, which in general is similar to AlphaFold_DB. On the randomly generated vRCC the maximum tends to be larger, indicating that Experimental_DB and AlphaFold_DB tend to include more similar vRCC (smaller angles) than those observed randomly. In all these cases, the parallelism or semi-parallelism is meager at 5° or less. At that angle, vRCC are more likely to share fewer types RCC, hence facilitating the testing of whether pairs of interacting proteins do share the same types of RCC, like in the case of homodimers. Alternatively, using 25° would not allow to note differences.

Consequently, the pairs of proteins with an angle $\theta < 5°$ were counted, allowing for the calculation of a parallelism frequency (see Figure 4.3, data and codes available at https://github.com/gdelrioifc/PPI-RCC2). We noticed that AlphaFold_DB has smaller frequency of parallels

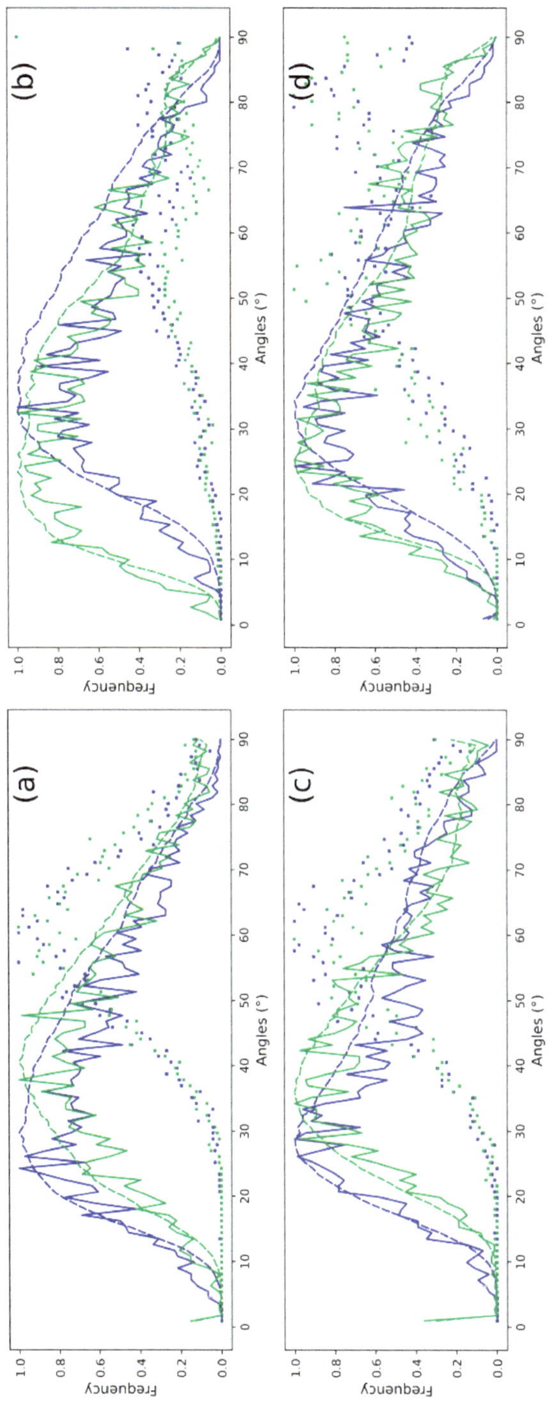

Figure 4.2 Angle frequency distributions of vRCC built using 7 Å with (a) or without (b) side chains, and for 8 Å with (c) or without (d) side chains. The distribution for Experimental_DB (green) and AlphaFold (blue) are shown. Interacting (continuous lines), non-interacting (dashed lines) and random (dotted) pairs are shown for each data set.

vRCC than the Experimental_DB in both, interacting and non-interacting proteins. This difference may be due to the use of AlphaFolds models without any filter (pLDDT or any other). In any case, the randomly simulated pairs of vRCC had again a lower likelihood of parallelism or semi-parallelism than Experimental_DB or AlphaFold_DB, indicating that despite the differences observed in these sets, both differ from random.

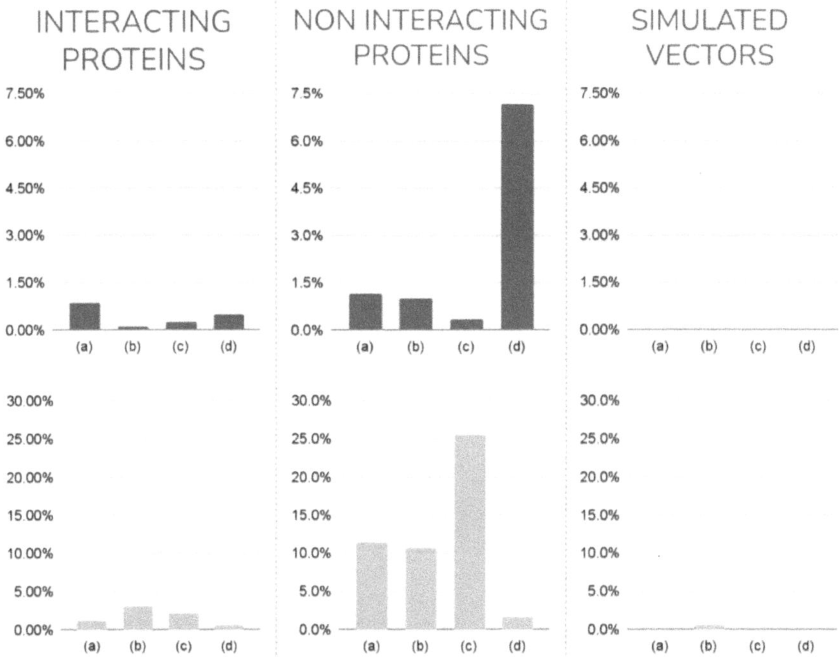

Figure 4.3 Frequency of parallelism in each database for 7 Å. (a) with and (b) without side chains, and for 8 Å (c) with and (d) without side chains. At the bottom (gray bars) Experimental_DB has vRCC for experimental structures for four organisms (*C. elegans, A. thaliana, H. sapiens,* and *S. cerevisiae*). At the top (dark gray) AlphaFold_DB has vRCC for predicted structures for the same four organisms. Interacting pairs were obtained from a public repository (Mosca et al. 2013) and the corresponding vRCC were obtained as previously reported (Fontove and Del Rio 2020). Non-interacting pairs were generated randomly considering that two random proteins rarely are going to interact with each other.

In summary, the vRCCs for proteins whose structure was determined experimentally or with AlphaFold exhibit parallelism relationships different from what would be expected randomly, yet AlphaFold models show a clearly different pattern of vRCC parallels from the experimentally determined protein structures; this is also observed in the angle distribution between AlphaFold models and experimentally determined protein structures. Interestingly, interacting proteins exhibit less parallelism than a sample of non-interacting proteins, indicating that vRCC from real proteins tend to accumulate parallels.

DISCUSSION

In this chapter, we have introduced a vector representation of the 3D structure of proteins that account for their densest regions, their maximal cliques, which are defined as RCC. We propose a general framework to describe how proteins may have evolved to fill in spaces by combining these maximal cliques either using a single protein monomer or multiple protein monomers. In material sciences it has been shown that packing density is related to the vibrational spectrum of materials (Hettiarachchi and Mampearachchi 2018, Li, C.X et al. 2020). In a similar fashion, packing features are expected to be related to the vibration modes of proteins (Rischel et al. 1998). In relation to the goals of our work to correlate the packing/vibration with the folding and protein binding, it has been shown that residues important for folding and PPI residues present high frequencies of vibration (Haliloglu et al. 2005), an observation consistent with the theoretical claims (Tsai et al. 1998). Therefore, RCC sets the study of protein structure and function within the framework of other physics-based phenomena.

To further support this framework, in this chapter we compared different PPI prediction methods with one previously reported by our group that we refer to as PPI-RCC. We claim that if maximal cliques are linked to the structure and function of proteins, this model should comply with Occam's razor criteria, that is to achieve the best or close to the best modeling efficiency using the smallest number of variables. First, we observed a large diversity in datasets used to train or evaluate PPI models (see Table 4.1). Among the top predictors, DeepPPI (97.4%) uses too many variables (>500). On the other hand, PPI-RCC uses some of the smallest number of variables (26) and present an average prediction efficiency close to DeepPPI (97%). PPI-RCC in the test set rendered a sensitivity and specificity values of 95.2 and 83.3%, respectively, while DeepPPI did not report sensitivity values, but reported a 94.5% in their specificity with their test sets. Both DeepPPI and PPI-RCC, used similar positive PPI data. For the negative instances of PPI, DeepPPI chose pairs of proteins that are not reported to localize in the same cellular compartment and are not reported to interact, while PPI-RCC use pairs included in the Negatome database, that is, pairs of proteins that are not in contact in PDB complexes and are not reported to interact. This difference may explain the observed efficiencies in classifying negative instances of PPI. Finally, DeepPPI use *S. cerevisiae* for training and other

5 different organisms to test its predictions, hence it is a reasonable set to test for the generalization of their model. PPI-RCC used data from protein complexes for training and initial testing; in the current work, we included a similar set of proteins used by DeepPPI and others (see Table S1) to further test PPI-RCC; this new set involves proteins that are known to interact, but their 3D structure were not solved in complex. Our results show that PPI-RCC was able to predict correctly almost all positive PPI in this new set, hence confirming the ability to generalize PPI classification. We conclude that PPI-RCC represent an efficient PPI classifier since it requires 20 times less variables to achieve similar efficiencies than DeepPPI, complying with the Occam's razor criteria.

If maximal cliques are used to fill in spaces, like in tessellations, we proposed that different PPI may have evolved by maintaining some maximal cliques and adding some new ones if any. For instance, many PPI are the result of two identical interacting protein monomers, hence this is a clear example that PPI are formed by maintaining some maximal cliques. To evaluate if this a trend among positive instances of PPI, we used three sets: Experimental_DB, AlphaFold_DB and Randome_DB (see above). We observed that Experimental_DB and AlphaFold_DB have larger frequencies of parallels or semi parallels ($<5°$) among the interacting proteins than Random_DB, supporting the divergent idea of tessellations. It is a striking result that AlphaFold models are clearly packed differently than real structures, thus it would be worth examining whether the use of different filters (e.g. pLDDT) may change this observation. That is beyond the goal of this study.

RCC account for the densest regions of proteins and here we propose a theoretical framework, tessellation divergence, to explain the evolution of protein structure. Such framework upon further development, may assist in identifying sets of residues that may form different maximal cliques, hence rendering IDR and metastable proteins. Thus, tessellations and RCC represents a novel framework to study the structure and function relationship of proteins.

In summary, we showed that a method based on counting maximal cliques (RCC) is among the most efficient tools to predict PPI, with an average reliability of 97%. We also showed that proteins represented as a vector of RCC (vRCC) tend to have more parallels or semi parallels than randomly generated vectors. Interestingly, non-filtered AlphaFold models present a different distribution of parallels and semi parallels than experimentally determined protein structures. These findings reveal properties of proteins and models previously unknown, highlighting the relevance of using RCC to further characterize and study protein structure and function.

ACKNOWLEDGEMENTS

Dra. Maria Teresa Lara Ortiz for her technical assistance in the preparation of the materials required for the development of this work.

REFERENCES

Alderson, T. Reid, I. Pritišanac, Đ. Kolarić, A.M. Moses and J.D. Forman-Kay. 2023. Systematic identification of conditionally folded intrinsically disordered regions by AlphaFold2. Proc. Natl. Acad. Sci. 120(44): e2304302120. doi:10.1073/pnas.2304302120.

Bagci, Z., R.L. Jernigan, and I. Bahar. 2002. Residue coordination in proteins conforms to the closest packing of spheres. Polymer. 43(2): 451–459. doi:10.1016/S0032-3861(01)00427-X.

Blohm, P., G. Frishman, P. Smialowski, F. Goebels, B. Wachinger, A. Ruepp, et al. 2014. Negatome 2.0: A database of non-interacting proteins derived by literature mining, manual annotation and protein structure analysis. Nucleic Acids Res. 42 (Database issue): D396–400. doi:10.1093/nar/gkt1079.

Bret, H., J. Gao, D.J. Zea, J. Andreani and R. Guerois. 2024. From interaction networks to interfaces, scanning intrinsically disordered regions using AlphaFold2. Nature Commun. 15(1): 597. doi:10.1038/s41467-023-44288-7.

Chakravarty, D., J.W. Schafer, E.A. Chen, J.F. Thole, L.A. Ronish, M. Lee, et al. 2024. AlphaFold predictions of fold-switched conformations are driven by structure memorization. Nat. Commun. 15(1): 7296. doi:10.1038/s41467-024-51801-z.

Corral-Corral, R., E. Chavez, and G. Del Rio. 2015. Machine learnable fold space representation based on residue cluster classes. Comput. Biol. Chem. 59 Pt A: 1–7. doi:10.1016/j.compbiolchem.2015.07.010.

Dill, K.A. 1985. Theory for the folding and stability of globular proteins. Biochemistry. 24(6): 1501–1509. doi:10.1021/bi00327a032.

Dishman, A.F. and B.F. Volkman. 2022. Design and discovery of metamorphic proteins. Curr. Opin. Struct. Biol. 74: 102380. doi:10.1016/j.sbi.2022.102380.

Dunker, A.K., J.D. Lawson, C.J. Brown, R.M. Williams, P. Romero, J.S. Oh, et al. 2001. Intrinsically disordered protein. J. Mol. Graphics Modell. 19(1): 26–59. doi:10.1016/S1093-3263(00)00138-8.

Dupuis, F., J.-F. Sadoc, R. Jullien, B. Angelov and J.-P. Mornon. 2005. Voro3D: 3D voronoi tessellations applied to protein structures. Bioinformatics. 21(8): 1715–1716. doi:10.1093/bioinformatics/bth365.

Evans, R., M. O'Neill, A. Pritzel, N. Antropova, A. Senior, T. Green, et al. 2022. Protein complex prediction with alphafold-multimer. bioRxiv. doi:10.1101/2021.10.04.463034.

Fontove, F. and G.D. Rio. 2020. Residue cluster classes: A unified protein representation for efficient structural and functional classification. Entropy. 22(4): 472. doi:10.3390/e22040472.

Gupta, Munishwar Nath and Vladimir N. Uversky. 2024. Protein structure–function continuum model: emerging nexuses between specificity, evolution, and structure. Protein Sci. 33(4): e4968. doi:10.1002/pro.4968.

Hales, T.C. 1997a. Sphere packings, I. Discrete Comput. Geom. 17(1): 1–51. doi:10.1007/BF02770863.

Hales, T.C. 1997b. Sphere packings, II. Discrete Comput. Geom. 18(2): 135–149. doi:10.1007/PL00009312.

Haliloglu, T., O. Keskin, B. Ma and R. Nussinov. 2005. How similar are protein folding and protein binding nuclei? Examination of vibrational motions of energy hot spots and conserved residues. Biophys. J. 88(3): 1552–1559. doi:10.1529/biophysj.104.051342.

Hettiarachchi, H.A. C.K. and W.K. Mampearachchi. 2018. Effect of vibration frequency, size ratio and large particle volume fraction on packing density of binary spherical mixtures. Powder Technol. 336: 150–160. doi:10.1016/j.powtec.2018.05.049.

Jumper, J., R. Evans, A. Pritzel, T. Green, M. Figurnov, O. Ronneberger, et al. 2021. Highly accurate protein structure prediction with alphafold. Nature. 596(7873): 583–589. doi:10.1038/s41586-021-03819-2.

Kewalramani, N., A. Emili and M. Crovella. 2024. State-of-the-art computational methods to predict protein–protein interactions with high accuracy and coverage. Proteomics. 23:(21–22): e2200292. doi:10.1002/pmic.202200292.

Kizilörenli, E. and F. Maden. 2021. Tessellation in architecture from past to present. IOP Conf. Ser. Mater. Sci. Eng. 1203(3): 032062. doi:10.1088/1757-899X/1203/3/032062.

Lara O., M. Teresa, V.M. García and G.D. Rio. 2021. Saturation mutagenesis of the transmembrane region of HokC in escherichia coli reveals its high tolerance to mutations. Int. J. Mol. Sci. 22(19): 10359. doi:10.3390/ijms221910359.

Lensink, M.F., G. Brysbaert, N. Nadzirin, S. Velankar, R.A.G. Chaleil, T. Gerguri, et al. 2019. Blind prediction of homo- and hetero-protein complexes: The CASP13-CAPRI experiment. Proteins. 87(12): 1200–1221. doi:10.1002/prot.25838.

Li, C.X, R.P. Zou, D. Pinson, A.B. Yu and Z.Y. Zhou. 2020. An experimental study of packing of ellipsoids under vibrations. Powder Technol. 361: 45–51. doi:10.1016/j.powtec.2019.10.115.

Lubell, D. 2009. A short proof of sperner's lemma. pp. 402. *In*: Gessel, I., Rota, GC. (eds) Classic Papers in Combinatorics. Modern Birkhäuser Classics. Birkhäuser Boston. doi:10.1007/978-0-8176-4842-8_28.

MacCallum, J.L., A. Perez and K.A. Dill. 2015. Determining protein structures by combining semireliable data with atomistic physical models by bayesian inference. Proc. Natl. Acad. Sci. 112(22): 6985–6990. doi:10.1073/pnas.1506788112.

Malagrinò, F., A. Diop, L. Pagano, C. Nardella, A. Toto and S. Gianni. 2022. Unveiling induced folding of intrinsically disordered proteins—protein engineering, frustration and emerging themes. Curr. Opin. Struct. Biol. 72: 153–160. doi:10.1016/j.sbi.2021.11.004.

Mosca, R., A. Céol and P. Aloy. 2013. Interactome3D: Adding structural details to protein networks. Nat. Methods. 10(1): 47–53. doi:10.1038/nmeth.2289.

Poot Velez, A.H., F. Fontove and G. Del Rio. 2020. Protein-protein interactions efficiently modeled by residue cluster classes. Int. J. Mol. Sci. 21(13): 4787. doi:10.3390/ijms21134787.

Porter, L.L., I. Artsimovitch and C.A. Ramírez-Sarmiento. 2024. Metamorphic proteins and how to find them. Curr. Opin. Struct. Biol. 86: 102807. doi:10.1016/j.sbi.2024.102807.

Rischel, C., D. Spiedel, J.P. Ridge, M.R. Jones, J. Breton, J.-C. Lambry, et al, 1998. Low frequency vibrational modes in proteins: changes induced by point-mutations in the protein-cofactor matrix of bacterial reaction centers. Proc. Natl. Acad. Sci. 95(21): 12306–12311.

Tsai, C. J., D. Xu, and R. Nussinov. 1998. Protein folding via binding and vice versa. Fold Des. 3(4): R71-80. doi:10.1016/S1359-0278(98)00032-7.

Tunyasuvunakool, K., J. Adler, Z. Wu, T. Green, M. Zielinski, A. Žídek, et al. 2021. Highly accurate protein structure prediction for the human proteome. Nature. 596(7873): 590–596. doi:10.1038/s41586-021-03828-1.

Uversky, V.N., J.R. Gillespie and A.L. Fink. 2000. Why are 'natively unfolded' proteins unstructured under physiologic conditions? Proteins. 41(3): 415–427. doi:10.1002/1097-0134(20001115)41:3<415::aid-prot130>3.0.co;2-7.

Varadi, M., S. Anyango, M. Deshpande, S. Nair, C. Natassia, G. Yordanova, et al. 2022. AlphaFold protein structure database: massively expanding the structural coverage of protein-sequence space with high-accuracy models. Nucleic Acids Res. 50(D1): D439–D444. doi:10.1093/nar/gkab1061.

Wang, X.-W., L. Madeddu, K. Spirohn, L. Martini, A. Fazzone, L. Becchetti, et al. 2023. Assessment of community efforts to advance network-based prediction of protein–protein interactions. Nat. Commun. 14(1): 1–14. doi:10.1038/s41467-023-37079-7.

Wright, P.E. and H. Jane Dyson. 1999. Intrinsically unstructured proteins: Re-assessing the protein structure-function paradigm. J. Mol. Biol. 293(2): 321–331. doi:10.1006/jmbi.1999.3110.

Xian, L. and Y. Wang. 2024. Advances in computational methods for protein–protein interaction prediction. Electronics. 13(6): 1059. doi:10.3390/electronics13061059.

Zhang, J., J. Durham and Q. Cong. 2024. Revolutionizing protein–protein interaction prediction with deep learning. Curr. Opin. Struct. Biol. 85: 102775. doi:10.1016/j.sbi.2024.102775.

Chapter **5**

Aromatic Clusters in Proteins: Folding, Drug Interactions and Protein-Protein Complexes

Esteban Lanzarotti[1], Lucas Defelipe[2], Marcelo Marti[3] and Adrian Turjanski*[3]

[1]Computer Science Institute, Faculty of Exact and Natural Sciences, University of Buenos Aires-CONICET, IQUIBICEN.

[2]European Molecular Biology Laboratory, Hamburg.

[3]Biological Chemistry Institute, Faculty of Exact and Natural Sciences, University of Buenos Aires-CONICET, IQUIBICEN.

INTRODUCTION

The rapidly increasing amount of protein structural information accumulated in the PDB provides an opportunity to view the structural space of biomolecules as a whole, to explore and understand the principles that govern structure-function relationships. Several studies on proteins have shown that interactions established by aromatic residues are important for the stabilization of protein structure, for the folding process, for the recognition of protein-protein interactions, and for ligand binding (Burley and Petsko 1985, Bashford et al. 1987,

*For Correspondence: aturjans@gmail.com

Serrano et al. 1991, Mitchell et al. 1994, Lesk and Fordham 1996, Chotia et al. 1998, Michnick and Shakhnovich 1998, de Araujo et al. 1999, Samanta et al. 1999, Kannan and Vishveshwara 2000, Larson and Davidson 2000, Bhattacharyya et al. 2002, Chelli et al. 2002, Aravinda et al. 2003, Johnson et al. 2007, Espinoza-Fonseca and García-Machorro 2008, Eidenschink et al. 2009). In fact, it was pointed out that "non-smelly" proteins, i.e., proteins that do not contain cysteine and aromatic residues (CFYWH-depleted), were virtually never found in PDB (Yan et al. 2020). Since "non-smelly" proteins typically did not independently fold into monomeric structures, instead being able to fold as a result of binding to interaction partners, it was concluded that structure of monomeric proteins is crucially dependent on the presence of cysteine and aromatic residues (Yan et al. 2020). To date, the dimeric interactions of aromatic rings have been extensively characterized and classified into three conformations depending on the orientation between the planar angles defined by the aromatic rings: face-to-face (or pi-stacking), face-to-edge (or T-shape), or edge-to-edge, taking into account that the assemblies also tend to adopt intermediate orientations. Previous studies of protein structure show that, of the three, the aromatic interaction is favored in the T-shape conformations (Burley and Petsko 1985, Hunter et al. 1991).

Aromatic molecules, in turn, tend to form larger clusters with an additive energetic nature, which also adopt specific conformations, as seen in the case of benzene clusters in vacuum (Kannan and Vishveshwara 2000, Easter et al. 2005, Morimoto et al. 2007, Gonzalez and Lim 2001, Dang 2000, Engkvist et al. 1999, De Meijere and Huisken 1990, Tauer and Sherrill 2005). These clusters have been studied both experimentally and computationally. These larger clusters also occur in protein structures and are important for protein structure and function. Aromatic rings are widely used in drug design because the introduction of aromatic interactions between a drug and its target can help optimize both affinity and specificity (Li et al. 2013). This popularity of aromatic rings is partly due to the fact that they allow the generation of common "skeletons" in lead compounds that can be optimized to improve selectivity according to the requirements of the target (Ritchie and Macdonald 2014). On the other hand, the lower the number of aromatic rings a drug contains, the more developable it is in terms of known candidates on the market. More specifically, having more than 3 aromatic rings makes the compound more likely not to become a good drug, increasing the risk of developing such a compound (Ritchie and Macdonald 2009). In addition, if a compound is known to have low solubility in the optimization process, it is often beneficial to reduce the number of aromatic rings. This suggests that aromatic rings are an important resource that should be considered and rationalized. Therefore, the position of aromatic rings in a drug should be optimized according to the most important interactions with its target. Regarding protein-protein interaction interfaces, aromatic residues are important in the recognition and binding process, playing the role of "anchor" residues (Rajamani et al. 2004). Protein interfaces are characterized by polar and aromatic residues near the center of the contact region, and only in 20% of the complexes aliphatic residues play the role of anchors. Furthermore, it has been seen (Ma et al. 2003) that conserved aromatic residues, mainly TRP, and to a lesser extent PHE and HIS, on the surface of proteins strongly indicate a protein-protein interaction interface.

AROMATIC CLUSTER DEFINITION

To geometrically characterize aromatic clusters and understand their role in proteins, they can be divided into: (i) those involved only in protein folding, namely Intraprotein (IP), i.e., taking into account aromatic interactions that occur between residues of the same protein chain, (ii) those present at sites where drug-like ligands are bound, called Protein-Drug (PD) and (iii) those clusters involved in Protein-Protein interaction interfaces (PP).

In order to recognize aromatic interactions, we must first clearly define what we call an aromatic ring. An aromatic ring is a group of 5 or 6 atoms, so in the case of fused systems such as indole, 2 rings are used: a benzene and a pyrrole. In proteins, one can typically find four natural aromatic residues, histidine, phenylalanine, tyrosine, and tryptophan. Once the rings have been identified, it is known that aromatic interactions can be described mainly by the distance between the centers of each ring (Chelli et al. 2002). However, some angles have also been proposed to describe, for example, the relative position of the rings with respect to each other, or the angle between the normals of the planes (Brocchieri and Karlin 1994, McGaughey et al. 1998, Marsili et al. 2008). The two main parameters are the distance (d) and the planar angle (α) between the planes defined by the aromatic rings. These two parameters are very useful to visualize the different conformations in which two aromatic rings can interact: pi-stacking and T-shape. These parameters can be used to study their distribution among protein structures in order to identify and analyze interactions based on previously defined sets of aromatic rings: IP, PD and PP.

Aromatic interactions define a relationship between two aromatic rings. We can think of these interactions as links in a network where the nodes are the aromatic rings. To characterize three or more rings that interact to form larger clusters, each aromatic cluster is defined as a set of the connected components of the interaction network (Lanzarotti et al. 2011). Using this criterion, the aromatic clusters present can be geometrically characterized with respect to the three types of aromatics previously defined: IP, PD and PP (Lanzarotti et al. 2020). For PD-type complexes, only those Drug-Residue (D-R) interactions that exist between the aromatic rings in the residues and the aromatic rings in the drugs are taken into account. In the case of PP, we are left with those Residue-Residue (R-R) interactions that involve at least one residue in each chain. For IP, only those R-R interactions between residues of the same polypeptide chain are considered. It should be noted that for both PD and PP, in order to count the number of interactions that the clusters have, it is important to count the R-R interactions established by the rings of the same protein chain, so as not to underestimate the most compact clusters.

AROMATIC INTERACTIONS IN THE PDB

As can be seen in Table 5.1, for PD and PP there are 22,676 and 21,603 aromatic interactions, respectively, among the analyzed data, while in the case of IP there are 467,775. As for the clusters, the trend is similar, except that for PP and PD an average of 1.82 and 2.19 interactions per cluster are obtained, respectively, while for IP there are 3.93 interactions per cluster, indicating that the IP clusters are more packed than the PD and PP clusters. It has also occurred that in the 3 ring sets, there are cases where not a single aromatic interaction can be found. The proportion of those that have at least one aromatic interaction shows differences, in IP it reaches up to 88%, while for PD and PP it is only 49 and 53% respectively.

Table 5.1 Aromatic cluster statistics. From the second to the fourth column, in the following order: Intraprotein (IP), Protein-Drug (PD), and Protein-Protein (PP). The percentages in the second row are relative to the total entries in the first row

	Intraprotein	**Protein-Drug**	**Protein-Protein**
Number of entries	18,547	17,352	8,283
Entries with at least one cluster	16,430 (88%)	8,560 (49%)	4,422 (53%)
Number of aromatic rings	118,821	10,328	11,827
Number of aromatic interactions	467,775	22,676	21,603

More than 65% of all drugs analyzed have at least one aromatic ring, with those with 2 aromatic rings being more common. Also, more than 86% of the compounds containing aromatic rings are involved in at least one aromatic cluster, and those forming only one cluster represent more than 50% of the cases (Lanzarotti et al. 2020). Regarding the quantity of different aromatic rings present between each of the drugs, it turns out that the most abundant ring is benzene, reaching 55.5% of all rings belonging to compounds, followed by pyrimidines (11.6%) and pyridines (8.1%). Examples of different aromatics rings in proteins and in drugs can be seen in Figure 5.1.

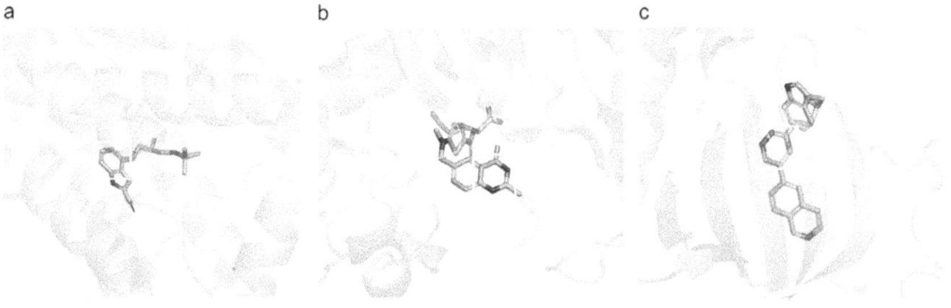

Figure 5.1 Examples of different aromatics rings in drugs interacting with proteins. Aromatic rings highlighted with yellow squares: (a) Benzene (b) Pyrrole (c) Pyridine.

Using the 3 types of clusters detected, the aromatic interactions involved can be analyzed, comparing them in terms of distance and planar angle distributions. As can be seen in Figure 5.2a, using the planar angle, the 2 energetic minima can be distinguished in the IP set: T-shape with a planar angle greater than 60° and pi-stacking with a planar angle less than 15°. In this case, a very slight enrichment is observed for the pi-stacking case, but in general there is no differential abundance between the two conformations and, furthermore, the planar angle distribution is more uniform than in the other two cases (PD and PP).

In the case of aromatic interactions at protein interaction interfaces, the distribution of interactions (Figure 5.2b) shows a surface with the same characteristics as in the previous case (IP). No preferred conformation between the two previously described conformations is observed. Both distributions coincide and no significant differences are observed, except for a very slight increase at 3.5 Å, which corresponds to the optimal value of the interaction distance in the pi-stacking conformation (McGaughey et al. 1998). However, in the case of PD interactions (Figure 5.2c), a more pronounced minimum is found at 3.5 Å compared to IP and PP. This is in agreement with the results obtained by (Chelli et al. 2002), where an increase in the difference of free energies of the pi-stacking conformation over the T-form is observed in hydrophilic environments. Aromatic interactions between drugs and proteins are more compact, in terms of the distance between the centers of mass of the rings, than in the case of inter- and intraprotein interactions. In the case of IP, this is not the case due to the hydrophobic nature of protein cores. With regards to PP clusters, the interactions appear to occur in an intermediate environment. This is a consequence of the fact that

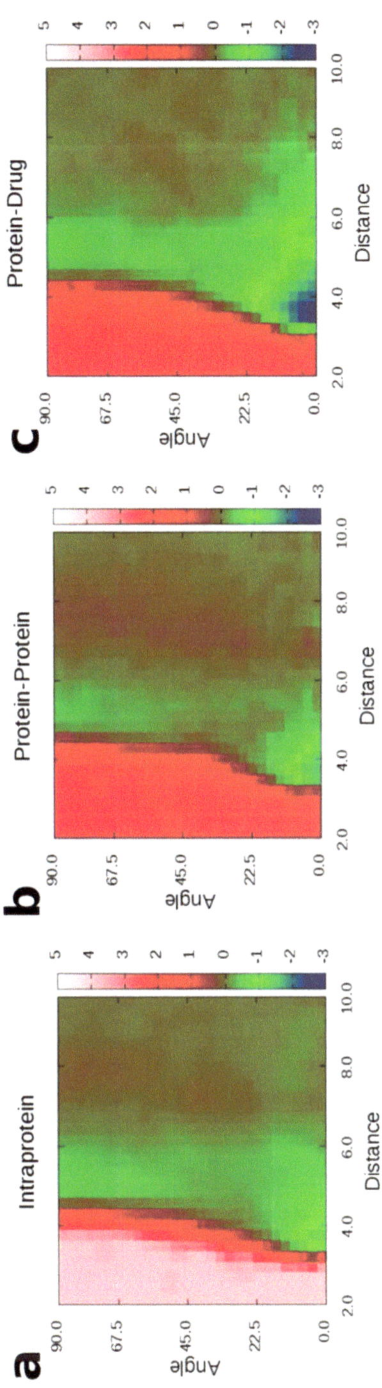

Figure 5.2 **Distribution of aromatic interactions.** Relationship between the aromatic planar angle and the distance between the center of mass of the rings. Two types of interactions can be distinguished: T-shape with angles greater than 60° and pi-stacking with angles less than 10°. (a) Intraprotein (IP), (b) Protein-Protein (PP), (c) Protein-Drug (PD).

the interfaces between proteins are slightly more hydrophilic than the protein cores, but not as much as the pockets where drug-like ligands bind (Giovambattista et al. 2008).

AROMATIC RING CLUSTERS

In general, Table 5.2 shows how the abundance of aromatic clusters decreases with increasing cluster size. In the case of IP, an exponential decrease is observed since the amounts of clusters of a given size is about half that of clusters with one less ring. The same is true for PD. For PP, however, this relationship is slightly lost and a more pronounced decrease is observed with increasing cluster size, indicating that larger clusters are less abundant in PP than in IP and PD. Dimeric clusters are more frequent in PP than in PD, reaching 58.6 and 47.5%, respectively. On the other hand, trimeric clusters reach 25.2% in PD and 23.3% in PP, tetramers 13.4 and 10.1%, thus the largest clusters are more common in PD than in PP. Another difference is that the average number of interactions (Table 5.3) for each cluster is higher in PP than in PD. This is a general trend, across the different cluster sizes, the average number of interactions is higher in PP than in PD. This implies that PP complexes tend to form more compact clusters.

Table 5.2 Number of clusters detected according to their size, for the 3 sets: PD, PP, and IP. The numbers in parentheses correspond to the percentage of the total in the column

Cluster size	Protein-Drug	Protein-Protein	Intraprotein
2	4,900 (47.5)	6,927 (58.6)	49,565 (41.7)
3	2,596 (25.2)	2,750 (23.3)	25,983 (21.9)
4	1,379 (13.4)	1,201 (10.1)	13,461 (11.3)
5	664 (6.4)	452 (3.8)	8,121 (6.8)
6	398 (3.9)	235 (2.0)	6,026 (5.1)
7	193 (1.9)	115 (1.0)	3,905 (3.3)
>8	186 (1.8)	147 (1.2)	11,760 (9.9)

Table 5.3 Average number of interactions according to the cluster size, for the 3 sets: PD, PP, and IP

Cluster size	Protein-Drug	Protein-Protein	Intraprotein
2	1	1	1
3	2.26	2.41	2.14
4	3.82	4.07	3.50
5	5.45	5.92	4.90
6	7.31	7.57	6.41
7	9.15	9.58	8.03
>8	11.17	11.18	9.71

Aromatic clusters are found in great abundance as part of protein folds. However, as shown in Table 5.3, aromatic rings can also form clusters in the regions of interaction with other proteins and with small drug-like ligands, which are schematized in Figure 5.3. Intraprotein aromatic clusters can have different degrees of packing, as seen in Figure 5.3a, which corresponds to a ladder-type trimer (LAD) with 2 interactions, and in Figure 5.3b, which corresponds to a symmetric-type trimer (SYM) with 3 interactions. Again, as in Figure 5.3c, it is shown a tetramer formed by aromatic residues that are contiguous in the protein sequence, creating a region with greater potential to attract aromatic rings that are farther apart in the sequence. Then, systems such as that in Figure 5.3d can also be found, where there are drug-like ligands bound to 2 aromatic rings in a protein, and that in Figure 5.3f, where a drug forms a larger cluster (hexamer). Finally, although less common, there are also occurrences of large aromatic clusters in protein interfaces, such as that in Figure 5.3e, which corresponds to a pentamer formed by 4 tyrosines and a phenylalanine.

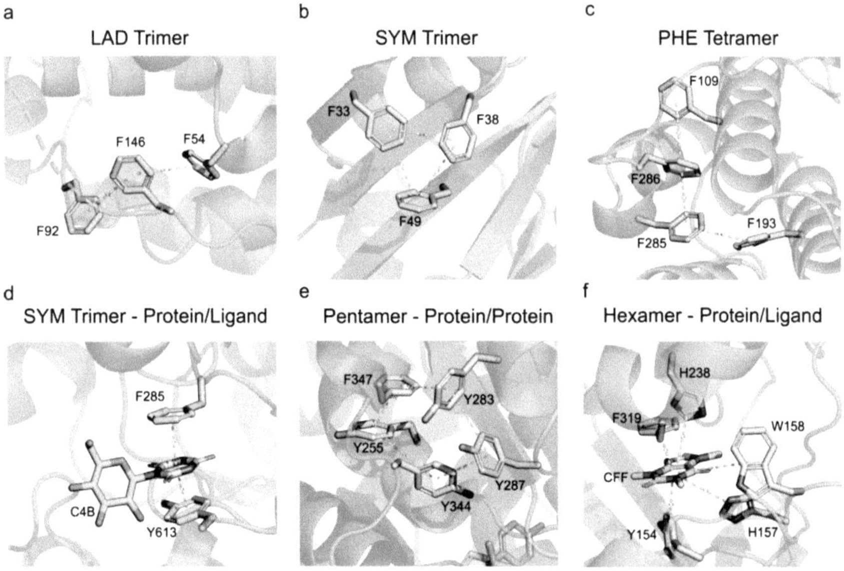

Figure 5.3 Intraprotein (IP), Protein-Drug (PD) and Protein-Protein (PP) aromatic clusters.

Composition of Aromatic Clusters

In the case of IP, the analysis of the amino acid composition of the trimers shows that the percentage of Phe is between 45 and 50%, for Tyr it is between 31 and 39% and for Trp it is between 15 and 17%, proportions that correspond to those perceived by the natural abundances of these amino acids. When analyzing the types of interactions in the different clusters, it can be calculated how many dimers are formed with each pair of amino acids. The corresponding results are: Phe-Tyr, 31%; Phe-Phe, 26%; Phe-Trp, 15%; Tyr-Tyr, 11%; Tyr-Trp, 9%; and Trp-Trp, 3%. The most frequent interactions are between Phe and Tyr, closely followed by those between Phe and Phe. Interestingly, Phe-Trp interactions are more frequent than Tyr-Tyr and Tyr-Trp interactions, probably due to the low degree of hydrophilicity that Tyr exhibits due to its polar OH group. Trp to Trp interactions occur with very low frequency, but this is a consequence of the low natural abundance of Trp in proteins, no more than 1.5% based on Uniprot.

For PD and PP, as shown in Tables 5.4 and 5.5, the distribution of amino acids by cluster size also follows the natural abundances, except that a slight preference for Trps can be observed in PP with respect to PD. This is in agreement with what has been demonstrated by (Ma et al. 2003) that exposed that Trps are usually a strong indication of a protein-protein interface. Another interesting fact is that as the cluster size increases, this preference of clusters in PP to have more Trps becomes more pronounced, which is due to the fact that each Trp is larger than

Table 5.4 Number of occurrences of the 4 residues in the clusters detected in protein-drug interaction regions. Trp, Bzn and Prl correspond to benzene and pyrrole, which form the indole system of tryptophan. The numbers in parentheses correspond to the percentage of the total for each column.

Residue	Dimers	Trimers	Tetramers	Pentamers
Phe	2,063 (42.1)	1,527 (37.2)	1,135 (36.0)	642 (32.6)
Tyr	1,541 (31.5)	1,207 (29.4)	809 (25.7)	462 (23.5)
His	989 (20.2)	622 (15.2)	493 (13.5)	262 (13.3)
Trp Bzn	215 (4.4)	465 (11.3)	425 (13.5)	346 (17.6)
Trp Prl	89 (1.8)	283 (6.9)	290 (9.2)	255 (13.0)

Table 5.5 Number of occurrences of the 4 residues in the clusters detected in protein-protein interaction regions. Trp, Bzn and Prl correspond to benzene and pyrrole, which form the indole system of tryptophan. The numbers in parentheses correspond to the percentage of the total for each column.

Residue	Dimers	Trimers	Tetramers	Pentamers
Phe	5,428 (39.2)	2,888 (35.0)	1,515 (31.5)	686 (30.3)
Tyr	4,470 (32.3)	2,030 (24.6)	1,050 (21.9)	457 (20.2)
His	3,022 (21.8)	1,225 (14.9)	692 (14.4)	286 (12.6)
Trp Bzn	682 (4.9)	1,130 (13.7)	825 (17.2)	438 (19.4)
Trp Prl	249 (1.8)	975 (11.8)	716 (14.9)	393 (17.4)

the rest of the amino acids in order to generate more contact surface and therefore participate in larger clusters.

Analysis of the Cluster Environment

In order to analyze the characteristics of the structural environment in which each cluster can be found inside the proteins, for each cluster in IP, it can be calculated, the average degree of hydrophobicity of the environment and the relative position with respect to the center of the protein. Then, for PD and PP, it is also possible to analyze the percentage of the exposed area and the percentage of contact presented by the rings of the residues that make up each cluster. The calculated values are:

- **Percentage exposed and percentage in contact:** On the one hand, it is possible to calculate the percentage of the exposed area for each residue, taking into account the rest of the atoms involved in the protein. It is calculated by dividing the solvent-accessible area of each residue in the context of the protein by the total surface area of the isolated residue. On the other hand, in the cases of PD and PP complexes, the percentage in contact of each residue is calculated as the surface value that remains hidden in the presence of the other interactor (the drug or another protein) with respect to the exposed area of the residue if the interactor were not present.
- **CCDR (Cluster to Center Distance Ratio):** This is the distance from the center of the cluster to the center of the protein relative to the overall size of the protein. It is calculated by dividing the distance from the center of mass of the cluster to the center of mass of the protein divided by the radius of gyration of the protein. Values less than 1 indicate that the cluster is located in the interior of the protein and values greater than 1 indicate that the cluster is located on the surface of the protein. Values close to 0 indicate that the cluster is located near the center of the protein.
- **EHI:** Environmental Hydrophobicity Index, calculated by averaging the value of the Hydrophobicity Index (Kyte and Doolittle 1982) over all residues interacting with the cluster. Interacting residues are defined as those residues whose side chain center of mass is less than 8A from the side chain of any aromatic residue among the residues in the cluster.

Regarding the percentage exposed (Figure 5.4), the results show that the clusters in the IP are exposed very little or rather buried in the protein structures, since the vast majority of them are exposed at most 10% and a very small percentage have values higher than 20%. Aromatic rings in PD clusters have a larger average exposed area than those in IP, reaching their maximum value in the distribution around 10% and surpassing the distribution of rings in IP from 5% exposure. This means that a randomly selected aromatic residue with an exposed percentage greater than 5–10% has more chances of being an interacting residue (with both drugs and other proteins) than a residue involved only in folding the protein structure. The clusters in PP have this tendency is even more pronounced, presenting a few cases less than 10% of exposure and exceeding the PD distribution from 20% of exposure. This means that the percentage of exposure of the aromatic rings is a parameter that explains a large part of the function that this ring fulfills in the protein structure.

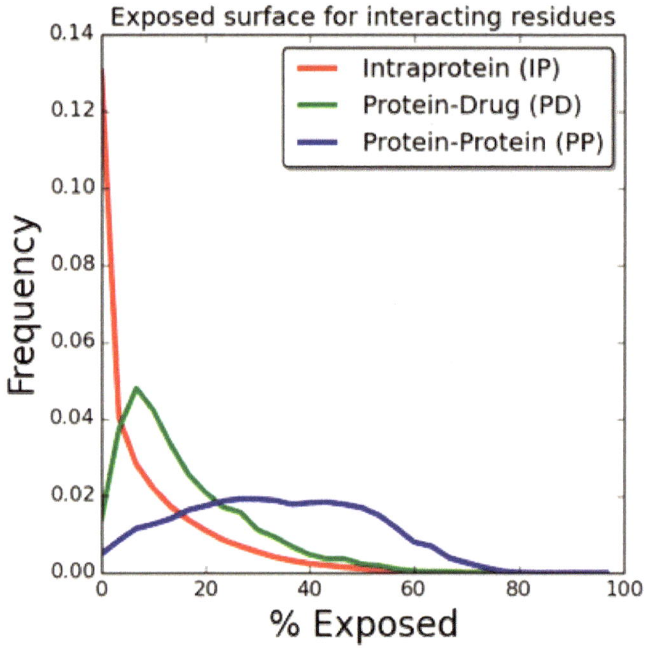

Figure 5.4 Solvent accessibility in aromatic clusters. Distribution of exposed surface percentage according to cluster type.

As can be seen in Figure 5.5A, most of the clusters are located in the interior of the protein (CCDR values < 1), showing that clusters in IP tend to be part of the hydrophobic core of the protein. These results are the same for ladder-type (LAD) and symmetric-type (SYM) trimers. In summary, both results show that, as expected, the clusters are buried inside the proteins and form part of their core. However, the interesting result is that the clusters in PD are located further inside the protein than in PP. Although this result can be expected because the rings in loose drugs can fit better into the interior of the protein than the rings in other proteins, it is interesting because the tendency is very pronounced. In order to analyze the nature of the environment of each cluster in IP, the EHI can be calculated for each of them, as previously defined, to measure the average degree of hydrophobicity of the surrounding neighboring residues. Negative values correspond to hydrophilic environments (the index for Arg is –4.5) and positive values to hydrophobic residues (the corresponding one for Leu is 4.5). Another fact to consider is that the average hydrophobicity, weighted by the abundance of the amino acids in the PDB, is -0.26. The results show that in general the environment of the aromatic clusters is rather hydrophobic, reaching an average EHI value of about 0.3, as can be seen in Figure 5.5B.

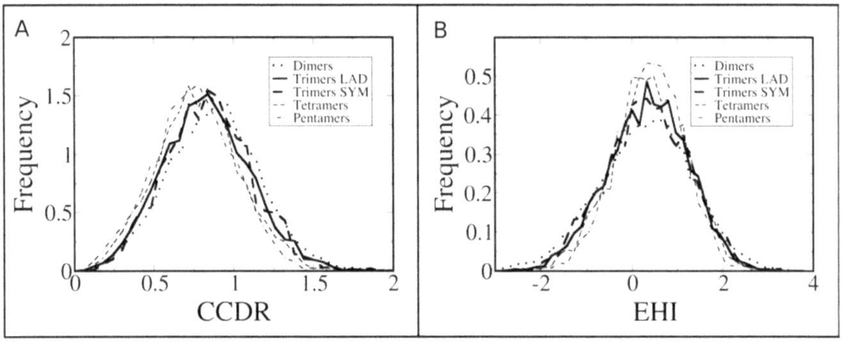

Figure 5.5 Analysis of the aromatic cluster environment in IP. (A) CCDR expresses the relative position of each cluster with respect to the protein core, (B) EHI expresses the average hydrophobicity of the residues in the environment of each cluster.

Now analyzing the clusters in PD and PP, in Figure 5.6 it can be seen that the relationship between the exposed surface of each residue that is part of the clusters and the contact surface with the interacting molecule, the other protein chain in the case of PP and the drug in the case of PP. It can be seen that the residues involved in the PP clusters are on average more exposed than the clusters in PD. This means that, given an aromatic residue, the exposed surface area provides information about the functional role played by this residue. In addition, the average percentage of contact is also higher in PP than in PD. Considering that the contact surface is one of the main parameters that allow modeling hydrophobic interactions (the greater the contact surface, the greater the exclusion of water molecules), this would indicate that the aromatic clusters in PP are more energetically stable than those found in PD.

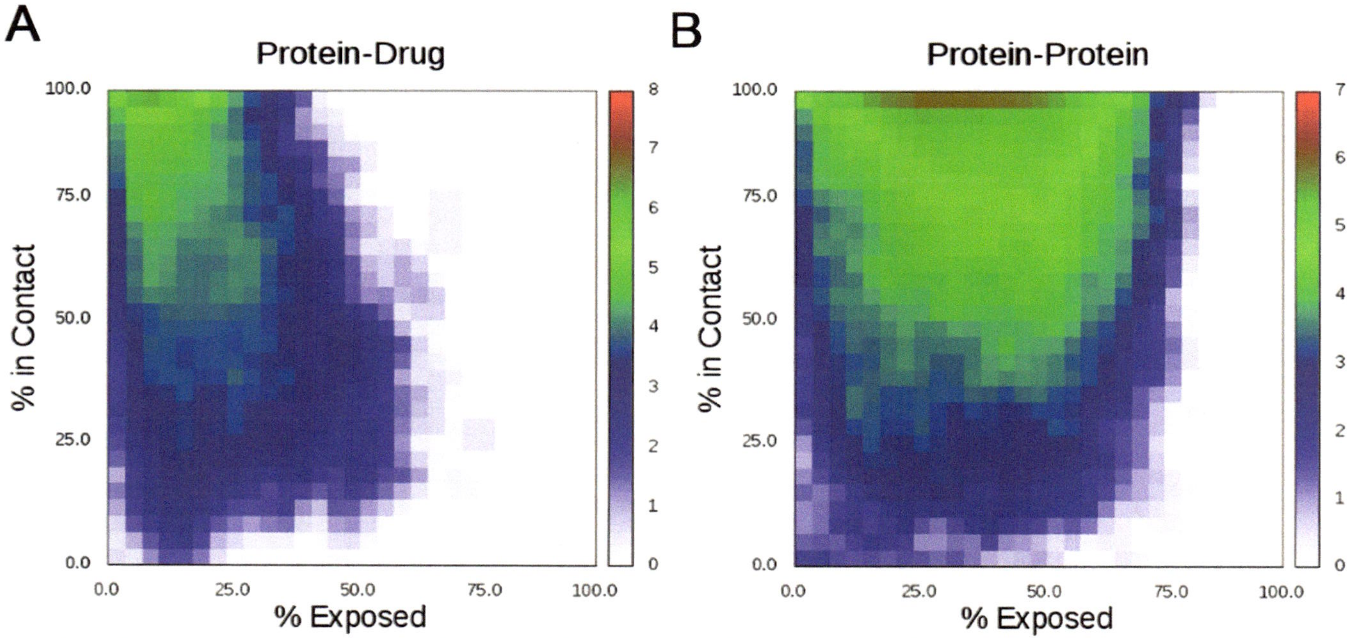

Figure 5.6 Percentage of Solvent Exposed Area and Percentage of Contact Area. (A) Percentage of exposed and contact area for protein-drug clusters, (B) Percentage of exposed contact area for protein-protein clusters.

CONCLUSION

In this chapter, it has been proposed that the idea of working with aromatic interactions beyond the dimer to a comprehensive exploration of the properties of their environment. Aromatic clusters can be divided into those that are part of the protein core (IP), those that interact with small ligands (PD), and those that are present at Protein-Protein interaction (PP) interfaces. Differences can be observed with respect to the environment in which they are located and the surface exposed to the solvent. Analysis of the occurrence of aromatic interactions in the structures deposited in the Protein Data Bank, focusing mainly on higher order interactions beyond the pure dimer has been carried out. It can be seen that the natural abundances of the interactions are consistent with the natural abundances of the amino acids. Finally, there are some differences in the interaction preference between PD and PP, while the former can adopt a traditional Pi-stacking interaction as well as T-shaped. The latter mostly prefers the traditional T-shape.

REFERENCES

Aravinda, S., N. Shamala, C. Das, A. Sriranjini, I.L. Karle and P. Balaram. 2003. Aromatic–aromatic interactions in crystal structures of helical peptide scaffolds containing projecting phenylalanine residues. J. Am. Chem. Soc. 125(18): 5308–5315.

Burley, S.K. and G.A. Petsko. 1985. Aromatic–aromatic interaction: a mechanism of protein structure stabilization. Science. 229(4708): 23–28.

Bashford, D., C. Chothia and A.M. Lesk. 1987. Determinants of a protein fold: Unique features of the globin amino acid sequences. J. Mol. Biol. 196(1): 199–216.

Bhattacharyya, R., U. Samanta and P. Chakrabarti. 2002. Aromatic–aromatic interactions in and around α-helices. Protein Eng. 15(2): 91–100.

Brocchieri, L. and S. Karlin. 1994. Geometry of interplanar residue contacts in protein structures. Proc. Natl. Acad. Sci. 91(20): 9297–9301.

Chelli, R., F.L. Gervasio, P. Procacci and V. Schettino. 2002. Stacking and T-shape competition in aromatic-aromatic amino acid interactions. J. Am. Chem. Soc. 124(21): 6133–6143.

Chothia, C., I. Gelfand and A. Kister. 1998. Structural determinants in the sequences of immunoglobulin variable domain. J. Mol. Biol. 278(2): 457–479.

Dang, L.X. 2000. Molecular dynamics study of benzene–benzene and benzene–potassium ion interactions using polarizable potential models. J. Chem. Phys. 113(1): 266–273.

de Araujo, A.F.P., T.C. Pochapsky and B. Joughin. 1999. Thermodynamics of interactions between amino acid side chains: experimental differentiation of aromatic-aromatic, aromatic-aliphatic, and aliphatic-aliphatic side-chain interactions in water. Biophys. J. 76(5): 2319–2328.

De Meijere, A. and F. Huisken. 1990. CO_2-laser induced photodissociation studies of size-selected small benzene clusters. J. Chem. Phys. 92(10): 5826–5834.

Easter, D.C., D.A. Terrell and J.A. Roof. 2005. Monte carlo studies of isomers, structures, and properties in Benzene–Cyclohexane Clusters: Computation strategy and application to the dimer and trimer, $(C_6H_6)(C_6H_{12})n$, $n= 1–2$. J. Phy. Chem. A. 109(4): 673–689.

Eidenschink, L.A., B.L. Kier and N.H. Andersen. 2009. Determinants of fold stabilizing aromatic-aromatic interactions in short peptides. *In*: Peptides for Youth: The Proceedings of the 20th American Peptide Symposium. 73–74.

Engkvist, O., P. Hobza, H.L Selzle and E.W. Schlag. 1999. Benzene trimer and benzene tetramer: structures and properties determined by the nonempirical model (NEMO) potential calibrated from the CCSD (T) benzene dimer energies. J. Chem. Phys. 110(12): 5758–5762.

Espinoza-Fonseca, L.M. and J. García-Machorro. 2008. Aromatic–aromatic interactions in the formation of the MDM2–p53 complex. Biochem. Biophys. Res. Commun. 370(4): 547–551.

Giovambattista, N., C.F. Lopez, P.J. Rossky and P.G. Debenedetti. 2008. Hydrophobicity of protein surfaces: Separating geometry from chemistry. Proc. Natl. Acad. Sci. 105(7): 2274–2279.

Gonzalez, C. and E.C. Lim. 2001. Ab initio study of the intermolecular interactions in small benzene clusters: The equilibrium structures of trimer, tetramer, and pentamer. J. Phys. Chem. A. 105(10): 1904–1908.

Hunter, C.A., J. Singh and J.M Thornton. 1991. π-π interactions: the geometry and energetics of phenylalanine-phenylalanine interactions in proteins. J. Mol. Biol. 218(4): 837–846.

Johnson, R.M., K. Hecht and C.M. Deber. 2007. Aromatic and cation–π interactions enhance helix—helix association in a membrane environment. Biochemistry. 46(32): 9208–9214.

Kannan, N. and S. Vishveshwara. 2000. Aromatic clusters: a determinant of thermal stability of thermophilic proteins. Protein Eng. 13(11): 753–761.

Krause, H., B. Ernstberger and H.J. Neusser. 1991. Binding energies of small benzene clusters. Chem. Phys. Lett. 184(5–6): 411–417.

Kyte, J. and R.F. Doolittle. 1982. A simple method for displaying the hydropathic character of a protein. J. Mol. Biol. 157(1): 105–132.

Lanzarotti, E., R.R. Biekofsky, D.A. Estrin, M.A. Marti and A.G. Turjanski. 2011. Aromatic-aromatic interactions in proteins: beyond the dimer. J. Chem. Inf. Model. 51(7): 1623–1633.

Lanzarotti, E., L.A. Defelipe, M.A. Marti and A.G. Turjanski. 2020. Aromatic clusters in protein–protein and protein–drug complexes. J. Cheminform. 12(1): 30.

Larson, S.M. and A.R. Davidson. 2000. The identification of conserved interactions within the SH3 domain by alignment of sequences and structures. Protein Sci. 9(11): 2170–2180.

Lesk, A.M. and W.D. Fordham. 1996. Conservation and variability in the structures of serine proteinases of the chymotrypsin family. J. Mol. Biol. 258(3): 501–537.

Li, S., Y. Xu, Q. Shen, X. Liu, J. Lu, Y. Chen, et al. 2013. Non-covalent interactions with aromatic rings: current understanding and implications for rational drug design. Curr. Pharm. Des. 19(36): 6522–6533.

Ma, B., T. Elkayam, H. Wolfson and R. Nussinov. 2003. Protein–protein interactions: structurally conserved residues distinguish between binding sites and exposed protein surfaces. Proc. Natl. Acad. Sci. 100(10): 5772–5777.

Marsili, S., R. Chelli, V. Schettino and P. Procacci. 2008. Thermodynamics of stacking interactions in proteins. Phys. Chem. Chem. Phys. 10(19): 2673–2685.

McGaughey, G.B., M. Gagné and A.K. Rappé. 1998. π-stacking interactions: alive and well in proteins. J. Biol. Chem. 273(25): 15458–15463.

Michnick, S.W. and E. Shakhnovich. 1998. A strategy for detecting the conservation of folding-nucleus residues in protein superfamilies. Fold. Des. 3(4): 239–251.

Mitchell, J.B., C.L. Nandi, I.K. McDonald, J.M. Thornton and S.L. Price. 1994. Amino/aromatic interactions in proteins: is the evidence stacked against hydrogen bonding? J. Mol. Biol. 239(2): 315–331.

Morimoto, T., H. Uno and H. Furuta. 2007. Benzene ring trimer interactions modulate supramolecular structures. Angew. Chem. 119(20): 3746–3749.

Rajamani, D., S. Thiel, S. Vajda and C.J. Camacho. 2004. Anchor residues in protein–protein interactions. Proc. Natl. Acad. Sci. 101(31): 11287–11292.

Ritchie, T.J. and S.J. Macdonald. 2009. The impact of aromatic ring count on compound developability–are too many aromatic rings a liability in drug design? Drug Discov. Today. 14(21–22): 1011–1020.

Ritchie, T.J. and S.J. Macdonald. 2014. Physicochemical descriptors of aromatic character and their use in drug discovery: Miniperspective. J. Med. Chem. 57(17): 7206–7215.

Samanta, U., D. Pal and P. Chakrabarti. 1999. Packing of aromatic rings against tryptophan residues in proteins. Acta Crystallogr., Sect. D: Biol. Crystallogr. 55(8): 1421–1427.

Serrano, L., M. Bycroft and A.R. Fersht. 1991. Aromatic-aromatic interactions and protein stability: Investigation by double-mutant cycles. J. Mol. Biol. 218(2): 465–475.

Tauer, T.P. and C.D. Sherrill. 2005. Beyond the benzene dimer: An investigation of the additivity of π–π interactions. J. Phys. Chem. A. 109(46): 10475–10478.

Yan, J., J. Cheng, L. Kurgan and V.N. Uversky. 2020. Structural and functional analysis of "non-smelly" proteins. Cell. Mol. Life Sci. 77: 2423–2440.

Chapter **6**

Cell Penetrating Peptides with Repeat Sequences of Amino Acids

Ngoc Hai Trieu Phong[1], Mohammed A. Kamal[2], Duy Binh Tran[1] and Tohru Yamada*[1,3]

[1]Department of Surgery, Division of Surgical Oncology, University of Illinois College of Medicine, Chicago, IL 60612, USA.

[2]College of Liberal Arts and Sciences, University of Illinois Chicago, IL 60607, USA.

[3]Richard and Loan Hill Department of Biomedical Engineering, University of Illinois College of Engineering, Chicago, IL 60607, USA.

INTRODUCTION

Current approaches for delivery of therapeutic and diagnostic agents to target cells/tissues are imperfect, being often subject to limitations and inaccuracies due to poor cellular uptake and penetration, lack of specificity and selectivity, and toxicity concerns (Ezike et al. 2023). Despite comprehensive investigations on drug delivery systems, these methods typically result in low cell specificity, insufficient efficacy, and often exhibit toxicity. All this clearly conflicts with, for example, the

*For Correspondence: tohru@uic.edu

primary objective for developing a drug delivery system for cancer, which is to selectively target cancerous cells and avoid the healthy cells (Ezike et al. 2023, Chehelgerdi et al. 2023). Among promising developments in this field are Cell-Penetrating Peptides (CPPs), a diverse group of short peptides with the unique ability to cross cell membranes without compromising their structural or functional integrity, which have been extensively explored for the delivery of bioactive molecules, therapeutics, and theragnostic agents into cells (Ouyang et al. 2022, Bottens and Yamada 2022). A significant subset of CPPs is characterized by the presence of repeat sequences of amino acids. These repetitive motifs play a crucial role in the cell-penetrating properties of these peptides. Repeat sequences, such as polyarginine or amphipathic repeats, can enhance membrane interactions, facilitate cellular uptake, and improve the efficiency of cargo delivery (Kalafatovic and Giralt 2017). The number, type, and arrangement of these repeat sequences can significantly influence the penetration mechanism, cellular localization, and overall effectiveness of the CPP. Understanding the role of these repeat sequences is fundamental to the rational design and optimization of novel CPPs for various applications. These attributes are especially advantageous for therapeutic macromolecules, which typically exhibit restricted permeability across biological membranes (Komin et al. 2017). Previous studies have demonstrated that CPPs are highly effective in facilitating the intracellular delivery of various bioactive molecules, such as oligonucleotides, anticancer drugs, bioimaging agents, nanoparticles, proteins, and other peptides. The vast potential of CPPs is further validated by ongoing studies and clinical trials, which highlight their success in both *in vitro* and *in vivo* applications (Sadeghian et al. 2022).

In this chapter, we will provide a comprehensive overview of the mechanisms and current applications of CPP-based approaches for drug delivery. We will also discuss the current research and developments in this field, as well as the future directions and challenges.

HISTORICAL CONTEXT OF CPPs

The concept of CPPs first emerged in the late 1980s, when researchers discovered that the Trans-Activator of Transcription (TAT) protein from the Human Immunodeficiency Virus (HIV-1) could enter cells autonomously (Frankel and Pabo 1988, Green and Loewenstein 1988). This observation led to the identification of a specific sequence within the TAT protein that was responsible for its cell-penetrating ability, marking the discovery of the first CPP, now commonly known as the TAT peptide. Joliot et al. (1991) identified another natural CPP, the Antennapedia homeodomain from *Drosophila*. This 60-amino acids long polypeptide could enter nerve cells, expanding the understanding of how naturally occurring peptides could facilitate cellular entry (Joliot et al. 1991). In 1994, Derossi and his group (1994) further refined this concept by isolating a specific 16-amino acid sequence from the third helix of the Antennapedia homeodomain, which they found to be sufficient for cellular uptake. This peptide, later named "penetratin", became one of the most well-known CPPs, highlighting the potential of small peptide sequences to cross cellular barriers and facilitate the delivery of therapeutic agents (Derossi et al. 1994).

Following these discoveries, synthetic CPPs have been developed, which offer more versatility and control over peptide design for therapeutic purposes. In 1997, Vivès et al. identified the minimal sequence of the HIV-1 TAT protein (residues 48-60) that was sufficient for cellular uptake, simplifying the TAT structure while preserving its cell-penetrating properties (Vivès et al. 1997). In 1997, MPG represented a major breakthrough because it could form non-covalent complexes with nucleic acids, allowing for more efficient and flexible delivery of genetic material into cells (Morris et al. 1997). Another innovation emerged with the creation of the Transportan peptide, a chimera of the neuropeptide galanin and wasp venom peptide mastoparan (Langel et al. 1996, Langel 2021). This hybrid peptide demonstrated enhanced cell penetration and further expanded the range of synthetic CPPs, showcasing the potential for combining natural and synthetic elements to develop multifunctional delivery systems. These advancements in synthetic CPPs revolutionized the field of drug delivery, offering new possibilities for delivering therapeutic agents, including nucleic acids, proteins, and small molecules, with increased efficiency and specificity.

In the early 2000s, research shifted toward understanding the exact mechanisms by which these peptides enter cells. Initial studies suggested that CPPs can enter cells via direct translocation across the cell membrane, a process thought to be energy-independent (Rydström et al. 2011, Maniti et al. 2012). However, this view was soon challenged. More detailed studies revealed that endocytosis, rather than direct membrane translocation, played a significant role in CPP uptake, particularly when they were conjugated with larger molecular cargoes, such as proteins or nucleic acids (Ruseska and Zimmer 2020, Richard et al. 2003). Besides, the endocytic pathways implicated in CPP uptake were identified, including clathrin-mediated endocytosis, caveolae-mediated endocytosis, and micropinocytosis (Conner and Schmid 2003, Peters et al. 2003, Botos et al. 2008). Furthermore, the importance of experimental conditions, such as temperature and peptide concentration, were recognized in determining which uptake mechanism predominated (Fretz et al. 2007). Since then, this more complete knowledge of how CPPs are internalized has helped scientists make better CPP-based delivery systems.

These systems can deliver drugs and genes in a way that works best for each specific therapeutic need.

In the late 2000s and early 2010s, the field of CPPs was rapidly expanded with both the variety of CPPs and their applications. More advanced CPP databases and prediction tools were developed to better understand these peptides (Gautam et al. 2012, Agrawal et al. 2016, Preto et al. 2024). This was coupled with the increased use of computational methods, such as molecular dynamics simulations, to study CPP-membrane interactions at a deeper molecular level (Herce and Garcia 2007, Herce et al. 2009). These advancements helped refine the design of CPPs and guided their use in more targeted and effective applications. During this period, the number of clinical trials involving CPP-based therapies increased significantly, indicating the growing interest in translating these peptides into practical treatments.

In recent years, there has also been significant exploration of CPP-nanoparticle conjugates which offer a more controlled and enhanced delivery of therapeutic agents (Spicer et al. 2018, Gessner and Neundorf 2020). The potential of CPPs has expanded into areas like cancer therapy, gene editing, and diagnostics, further demonstrating their versatility. Despite these promising developments, efforts to bring CPP-based therapies to market are ongoing.

SOURCE OF CPPs

CPPs have been isolated from a wide range of sources, underscoring the widespread occurrence of cell-penetrating mechanisms in nature. Some of the well-known CPPs from viral sources, such as TAT peptide from HIV-1 (Frankel and Pabo 1988, Green and Loewenstein 1988, Qi et al. 2011) and VP22 from Herpes Simplex Virus type 1 (Elliott and O'Hare 1997, Tanaka et al. 2012), exemplify how viruses, with their evolved strategies for cellular entry, can be rich sources of CPPs. Additionally, antimicrobial peptides, including Magainin from the African clawed frog (Hirano et al. 2023) and LL-37 (Zhang et al. 2010), originally identified for their antimicrobial properties, have proven to be effective CPPs due to their inherent ability to traverse cell membranes. Venom peptides also contribute to this repertoire, with Maurocalcine from scorpion venom (Poillot et al. 2012) and Crotamine from rattlesnake venom (Campeiro et al. 2019, Rádis-Baptista 2021) showcasing potent cell-penetrating capabilities alongside their venomous roles.

Beyond natural sources, the field has seen the development of synthetic CPPs, such as polyarginine (Oh et al. 2014, Lunde et al. 2023), engineered to replicate the cell-penetrating properties of natural peptides, and the Model Amphipathic Peptide (MAP), which is another significant synthetic CPP. MAP is engineered to have a highly amphipathic α-helical shape. The amphipathicity was determined to be crucial for cellular uptake; a non-amphipathic analog with identical amino acids exhibited little uptake (Oehlke et al. 1998, Regberg et al. 2014). Furthermore, chimeric peptides, such as Transportan are created by merging sequences from different natural peptides, enhancing their functional versatility (Pooga et al. 1998, Bennett et al. 2024). Recent advances in discovery methods have facilitated the identification of novel CPPs with tailored properties.

TYPES OF CPPs

There are several types of CPPs, each offering valuable insights into their unique mechanisms of action. In this section, we focus on categorizing CPPs based on their physical, and chemical characteristics, with a primary emphasis on their charge properties. This classification allows us to better understand how different types of CPPs interact with cellular membranes and enhance their delivery potential, providing a clear framework for exploring their diverse biological applications.

Cationic CPPs

Cationic CPPs are distinguished by their abundance of positively charged amino acids, particularly arginine (R) and lysine (K), which give them a positive charge at physiological pH. This group of peptides includes TAT peptide (YGRKKRRQRRR), polyarginine peptides, such as R8 (RRRRRRRR) and R9 (RRRRRRRRR), and penetratin (RQIKIWFQNRRMKWKK). The main mechanism by which cationic CPPs facilitate cellular entry is through electrostatic interactions with negatively charged components of the cell membrane (Gori et al. 2023). The advantages of cationic CPPs due to their high positive charge enable them to deliver a wide range of cargo into cells, making them versatile tools in drug delivery and molecular research (Khairkhah et al. 2023). However, there are limitations to their use, such as potential toxicity at high concentrations. For example, when the number of repeating arginine residues exceeds 8 to 10, the CPPs may still be able to penetrate the cell membrane, but they might also cause membrane damage (Verdurmen and Brock 2011, Bottens and Yamada 2022). Cationic CPPs are also susceptible to enzymatic degradation, primarily due to their peptide nature and specific characteristics (Lucana et al. 2021).

Amphipathic CPPs

Amphipathic CPPs are the major classes of CPPs (more than 40% of the known CPPs are amphipathic) and are characterized by their distinct structural feature of having both hydrophobic and hydrophilic regions, which enables them to interact with different parts of the cell membrane. Due to their amphipathicity, such CPPs can undergo direct internalization regardless of their concentration (Singh et al. 2018). However, at higher concentrations, their membrane-disruptive ability can pose risks, potentially leading to cytotoxicity. For instance, MAP showed severe cytotoxicity at concentrations over 2 μM as established by conducting a fluorescein leakage assay (Scheller et al. 1999, Derakhshankhah and Jafari 2018). Transportan 10 (TP10) can induce significant membrane disruption as exposure to 10 μM of this CPPs resulted in approximately 40% leakage of lactate dehydrogenase (LDH) from human cancer cell lines (K562 and MDA-MB-231), indicating notable cytotoxicity (Saar et al. 2005). It has been observed that amphipathic CPPs compromise membrane integrity, triggering cellular responses such as elevated Ca^{2+} levels, membrane blebbing, and lysosomal exocytosis, whereas non-amphipathic CPPs do not (Lorents et al. 2012). The Model Amphipathic Peptide (MAP: KLALKLALKALKAALKLA) is a synthetic peptide that can penetrate the membrane by both endocytic and direct penetration. The MAP-cargo conjugate also showed much promise in cancer studies conducted *in vitro* (Silva et al. 2022).

Additionally, based on the sequence and lipid-binding properties of CPPs, they can be further categorized into primary and secondary amphipathic peptides, depending on their amino acid sequence and how they interact with phospholipid membranes. Primary amphipathic peptides are typically made up of more than 20 amino acids and have alternating hydrophobic and hydrophilic amino acids. Examples of primary amphipathic peptides include MPG (GALFLGFLGAAGSTMGAW SQPKKKRKV), CADY (Ac-GLWRALWRLLRSLWRLLWRA-Cya), and pVec (LLIILRRRIRKQAHAHSK) (Tripathi et al. 2018, Bottens and Yamada 2022). Secondary amphipathic peptides usually comprise fewer than 20 amino acids and can take on an α-helix or β-sheet conformation when they interact with phospholipid membranes. Secondary α-helical CPPs have a hydrophobic patch on one face, and the other face can be polar, anionic, or cationic. Secondary β-sheet CPPs have one hydrophobic stretch of amino acids and one hydrophilic stretch exposed to the solvent (Deshayes et al. 2006, Fernández-Carneado et al. 2004). MAP is an example of amphipathic peptide with an α-helical conformation (Silva et al. 2022). It is important to note that some CPPs can undergo conformational changes when interacting with membranes. For instance, many CPPs have a random conformation in solutions but can form certain secondary structures in hydrophobic environments or when interacting with membranes (Gong et al. 2018). The ability to form secondary structures (e.g., α-helices or β-sheets) upon membrane interaction can enhance the penetration ability of CPPs (Kalafatovic and Giralt 2017).

Hydrophobic CPPs

Hydrophobic CPPs are composed primarily of nonpolar amino acids, resulting in a low net charge. Several natural hydrophobic CPPs have been identified including the C105Y (CSIPPEVKFNKPFVYLI) peptide and its C-terminal domain PFVYLI as well as the Pep-7 peptide (SDLWEMMMVSLACQY) (Rhee and Davis 2006, Gao et al. 2002). Recently, several synthetic hydrophobic CPPs such as PFVYLI, a hexapeptide derived from the C-terminal portion of α1-antitrypsin were developed. It has reduced cytotoxicity and can localize to the nucleus (Mello et al. 2021); TP10 is a hydrophobic CPP that can be used to develop stapled hydrophobic CPPs (Tsuchiya et al. 2023). These peptides primarily interact with the hydrophobic core of the cell membrane's lipid bilayer, allowing them to enter cells via direct translocation or membrane fusion. Compared to cationic CPPs, hydrophobic CPPs generally exhibit lower toxicity (Mello et al. 2021), which may make them more favorable in some therapeutic applications.

CPPs with Repeat Sequences of Amino Acid

Certain CPPs are characterized by repeat sequences of amino acids (Table 6.1), making them highly effective for intracellular delivery. For example, polyarginine CPPs, [e.g., R8 and R9], consist of consecutive arginine residues, typically ranging from 6 to 12 amino acids, with the optimal number for cellular penetration being between 8 and 12. Although polyarginines generally do not have cell entry specificity, they demonstrate exceptional cellular uptake efficiency, often surpassing that of other CPPs. The high uptake efficiency of polyarginines is primarily due to the guanidinium groups on arginine residues, which play a critical role in interacting with the cell membrane. Arginine-rich CPPs use their guanidinium groups to bind anionic groups such as phosphates, carboxylates, and sulfates on the cell surface, facilitating membrane penetration through both direct translocation and endocytic pathways. These interactions also allow CPPs to partition into lipid glycerol regions, promoting deeper membrane penetration by forming transient pores. The flexibility in their mechanism makes arginine-rich CPPs efficient carriers for intracellular delivery (Hao et al. 2022). As drug

Table 6.1 List of CPPs with repeat sequences of amino acids

CPPs	Type of CPPs	Sequence	Isoelectric point (pI)*	No. of a.a	Molecular weight (kDa)	References
TAT	Cationic	GRKKRRQRRRPQ	13.3	12	1.62	(Frankel and Pabo 1988)
R8	Cationic	RRRRRRRR	13.5	8	1.27	(Hao et al. 2022)
R9	Cationic	RRRRRRRRR	13.6	9	1.42	(Hao et al. 2022)
MAP	Amphipathic	KLALKLALKALKAALKLA	11.2	18	1.88	(Silva et al. 2022)
WK4	Amphipathic	WKWKWKWK	11.0	8	1.27	(Mohammadi et al. 2015)
WK5	Amphipathic	WKWKWKWKWK	11.2	10	1.59	(Mohammadi et al. 2015)
GALA	Amphipathic	WEAALAEALAEALAEHLAEALAEALEALAA	3.3	30	3.03	(W. Li et al. 2004)
A2-17	Amphipathic	LRKLRKRLLRLWKLRKR	13.3	17	2.33	(Ohgita et al. 2019)
RR-22	Amphipathic	Ac-RGDGPLGLAGI3GR8-NH2	13.6	22	2.60	(Wang et al. 2020)

*Isoelectric point (pI) values were calculated by the online program (https://www.bachem.com/knowledge-center/peptide-calculator/)

delivery vehicles, CPPs with polyarginine improve siRNA delivery by enhancing membrane penetration and increasing bioavailability. These peptides also help protect siRNA and promote its endosomal escape, making them valuable for gene silencing therapies (Holjencin et al. 2021, Smidt et al. 2024).

Polyarginine CPPs also enhance the delivery of anti-cancer drugs by improving cell targeting and membrane penetration. They have been successfully applied in therapies for cervical cancer (Peng et al. 2022), ovarian cancer (Samec et al. 2022), colorectal cancer (Ser et al. 2020, Singh et al. 2021), lung cancer (Niazi et al. 2021), breast cancer (Yue et al. 2020), prostate cancer (Longoria-García et al. 2022, Zhou et al. 2010), and pancreatic cancer (Zhong et al. 2018). Moreover, to overcome the lack of specificity, arginine-rich CPPs are modified with tumor-targeting motifs, such as RGD (Wang et al. 2020, Chen et al. 2022), or conjugated with functional groups (Liu et al. 2017). These modifications improve tumor targeting and penetration, broadening their application in drug delivery systems.

Polyarginine CPPs are used in probes due to their excellent cell uptake capabilities. They enhance the internalization and functionality of probes in cells. These peptides have been used to develop fluorescent probes for whole-cell pH mapping (Xia et al. 2018, 2019), and probes for multicellular tumor spheroid (Gkika et al. 2021). Polyarginine peptides can enhance antimicrobial activity when conjugated with antimicrobial peptides (AMPs). These CPP-AMP conjugates have improved penetration, expanding the potential of repeated arginine peptides in antibiotic development and treatment. Conjugating R9 to AMPs (magainin and M15) significantly enhances their antimicrobial activity, especially against Gram-negative bacteria, by enabling cellular entry and DNA targeting rather than simply disrupting the bacterial membrane (Lee et al. 2019a). Polyarginine peptides are capable of crossing the Blood-Brain Barrier (BBB), making them promising candidates for delivering drugs to the brain. Studies have shown that CPPs, especially when combined with drugs or nanoparticles, can effectively transport therapeutic compounds across the BBB for treating brain diseases such as brain cancer.

While polyarginine peptides are the well-studied CPPs, others with repetitive sequences such as polyhistidine peptides also demonstrated cell-penetrating properties. Notably, histidine residues have a unique property due to their imidazole ring, which is neutral around physiological pH but becomes protonated in acidic environments (like tumor microenvironments). This protonation of the histidine residues in polyhistidine CPPs enhances their positive charge in the acidic tumor microenvironment, which increases their interaction with the negatively charged cell membranes. As a result, polyhistidine-rich CPPs can enter tumor cells more easily under these acidic conditions, making them especially useful for targeted drug delivery (Yao et al. 2017, Lee et al. 2019b). A study explores the use of gold nanoparticles (AuNPs) as agents in photothermal therapy for cancer treatment. Polyhistidine peptide was used as a drug delivery carrier to facilitate the internalization of AuNPs into tumor cells. The CPP-AuNPs demonstrated efficient uptake in RERF-LC-AI human lung squamous cancer cells, localizing primarily to intracellular lysosomes (Hori et al. 2022).

In another study, a modified version of L6 (RRWQWR), named HL6 (CHHHHHRRWQWRHHHHHC), which contains additional polyhistidine sequences and mono-cysteine on both ends of the original L6 peptide. These Cys residues can stabilize the structure through the formation of disulfide bonds (Lo and Wang 2008). By adding short polyhistidine peptides and mono-cysteine to L6 significantly enhances its ability to enter human bronchoalveolar carcinoma A549 cells. Unlike the original L6, which mainly uses endocytosis for cell entry, HL6 penetrates cells via direct membrane translocation—a more efficient mechanism for intracellular delivery. Furthermore, neither HL6 nor its complexes with quantum dots (HL6/QD) show little cytotoxicity to human bronchoalveolar carcinoma A549 cells or HeLa cells. They also proved that polyhistidine peptides play a key role in improving the CPP-mediated delivery process without harming the cells (Lee et al. 2019b). In addition, GALA is a synthetic peptide made up of 30 amino acids, featuring a repeating sequence of glutamic acid-alanine-leucine-alanine (EALA). It includes histidine and tryptophan residues, which serve as spectroscopic probes. Depending on pH of the solution, GALA disrupts membrane structures. This can be used as a model for making delivery systems for therapeutic molecules that can adapt to different environments (Li et al. 2004).

One of the main advantages of CPPs with repeat sequences is their enhanced cellular uptake compared to more complex peptide sequences. Additionally, these peptides are relatively simple to synthesize and modify, making them attractive for research and therapeutic applications. However, the simplicity of their structure can often result in reduced specificity, meaning they may not always target specific cell types or tissues effectively, which limits their use in certain contexts. Despite these challenges, CPPs with repeat sequences remain valuable tools for intracellular delivery due to their efficiency and versatility.

UPTAKE MECHANISM OF CPPs

The mechanisms underlying the cellular uptake of CPPs remain a subject of ongoing debate and are not fully understood yet. It is widely accepted that multiple pathways may be involved in the uptake of a single CPP, depending on various factors (Figure 6.1A). These factors include the sequence and concentration of the CPP, the properties

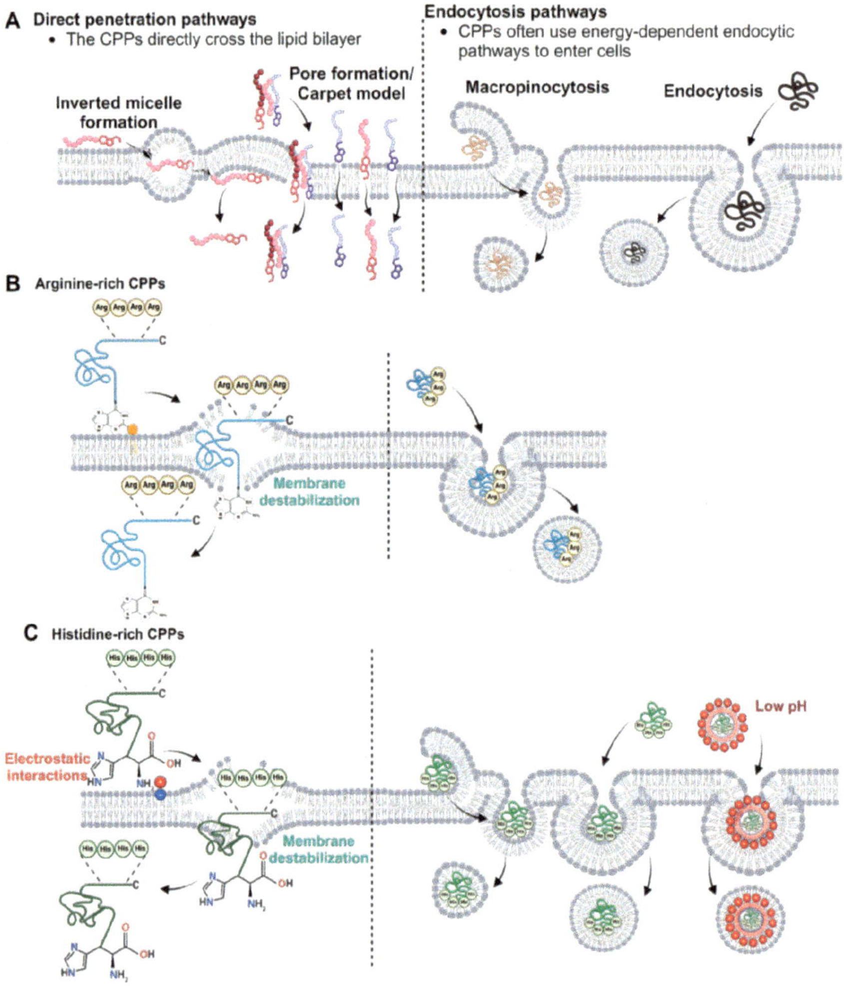

Figure 6.1 Schematic representation of major internalization pathways of cell-penetrating peptides. (A) Cellular uptake mechanisms of CPPs, including direct translocation and endocytosis, are illustrated. (B) for arginine-rich CPPs, the direct penetration pathway involves guanidinium groups forming bidentate bonds with negatively charged cell surface components, facilitating membrane disruption. (C) Histidine-rich CPPs penetrate membranes via two main mechanisms: Direct penetration, where electrostatic and hydrophobic interactions destabilize the membrane, and endocytosis, involving receptor-independent uptake and protonation within endosomes that promotes membrane disruption. The proton sponge effect enhances water and ion influx, leading to endosomal swelling and rupture, allowing the release of the peptide and its cargo into the cytoplasm.

of the attached cargo, the specific type of cell being targeted, and experimental conditions. For instance, arginine-rich CPPs (such as TAT, Penetratin, and oligoarginine peptides) utilize two pathways for cellular entry, endocytosis was the dominant route in most cases, while direct translocation can occur under certain conditions (Takeuchi and Futaki 2016) (Figure 6.1B). In the case of histidine-rich CPPs, the initial electrostatic interactions between positively charged CPPs and negatively charged cell surface components are essential for both direct penetration and endocytosis (Figure 6.1C). Without this attraction, CPPs would not get close enough to the cell surface to trigger either mechanism. CPPs with a higher positive charge are typically more efficient at entering cells (Ramaker et al. 2018). In cases of endocytic uptake, the major challenge is ensuring that the cargo escapes from endosomes once inside the cell, as an effective endosomal escape is necessary to prevent degradation and allow the cargo to reach its intended intracellular target. By adding peptides capable of disrupting membranes at acidic pH, they can enhance the endosomal escape of CPP-cargo conjugates (El-Sayed et al. 2009, Erazo-Oliveras et al. 2012). Additionally, CPPs can induce the budding and collapse of CPP-enriched vesicles from the endosomal membrane. In this process, portions of the endosomal membrane bud off to form vesicles, which then collapse, releasing their contents into the cytosol (Sahni et al. 2020). These insights are valuable for improving the efficiency of CPP-mediated delivery. The main mechanisms of cellular uptake for CPPs can be broadly categorized into two types: energy-independent and energy-dependent.

Energy-independent Direct Penetration

Direct penetration of CPPs across cell membranes is an energy-independent process that can occur through various mechanisms. One of the key mechanisms by which they do this is through the formation of transient pores in the lipid bilayer. The main mechanisms of CPP-mediated pore formation included: Carpet, Barrel-Stave, toroidal, inverted micelle models.

In the Carpet model, CPPs accumulate on the membrane surface, parallel to the lipid headgroups. At high concentrations, the CPPs disrupt the lipid packing and induce the formation of transient pores (Mudhakir and Harashima 2009). In the Barrel-Stave model, CPPs insert perpendicularly into the membrane and aggregate to form a barrel-like structure. The hydrophilic surfaces of the peptides face inward, creating an aqueous pore, while the hydrophobic residues interact with the lipid bilayer. In contrast, the toroidal model involves CPPs inducing the lipid bilayer to bend continuously, connecting the outer and inner leaflets of the membrane. The resulting pore is lined by both the peptide molecules and the lipid headgroups, creating a hydrophilic channel that involves both peptides and lipids. These models are typically associated with amphipathic peptides (Ruseska and Zimmer 2020). The inverted micelle model of CPP translocation begins with electrostatic interactions between the positively charged CPPs and negatively charged lipids on the cell surface. This attraction allows the peptides to anchor to the membrane, initiating membrane deformation. As a result, an inverted micelle forms, encapsulating both the CPPs and their cargo within the hydrophilic core of the lipid bilayer. This micelle then translocates through the membrane, effectively bypassing the need for energy-dependent processes. Once inside the cell, the micelle becomes destabilized, leading to the release of the CPPs and their cargo into the cytoplasm. This step is crucial for delivering the cargo to its intracellular target, ensuring the effectiveness of the CPP-mediated delivery system (Derakhshankhah and Jafari 2018). The inverted micelle model is more suited for hydrophobic CPPs (which rely on lipid interactions) (Bottens and Yamada 2022). While it can translocate small hydrophilic peptides, it is not efficient for highly cationic CPPs like TAT, which require a different mechanism (Derakhshankhah and Jafari 2018)

Energy-dependent Penetration

Energy-dependent penetration primarily occurs through endocytosis, which can be categorized into two main pathways: phagocytosis and pinocytosis. Phagocytosis is typically utilized by specialized cells such as macrophages, monocytes, and neutrophils to engulf large particles. In contrast, CPPs, being small peptides, are generally not internalized via phagocytosis. Instead, pinocytosis-particularly its subtypes such as macropinocytosis and clathrin-mediated endocytosis, caveolae-mediated endocytosis serves as the primary endocytic pathway for the internalization of CPPs (Ruseska and Zimmer 2020).

Endosomal Escape

Effective intracellular delivery of cargo using CPPs also depends on their ability to escape from endocytic organelles, primarily endosomes, and reach the cytosol. Poor endosomal release is often a major limitation, especially when large numbers of cargo molecules are required to trigger a biological effect.

Several studies suggested a possible mechanism involving membrane disruption. In this model, CPPs like the TAT peptide escape by making the endosome's membranes leaky, especially those rich in a lipid called BMP (bis(monoacylglycero) phosphate). This allows the CPPs and their cargo to leak out of the endosome and into the cell. These processes were influenced by the presence of BMP and occurred more efficiently at the acidic pH (Yang et al. 2010). The other study demonstrated that cyclic CPPs escape from endosomes by inducing vesicle budding from the endosomal membrane, followed by vesicle collapse and disintegration, which releases their cargo into the cytosol. This mechanism is facilitated by acidic conditions inside the endosome and is highly efficient for delivering larger molecules like proteins or nanoparticles (Sahni et al. 2020). Photochemical Internalization (PCI) is a new method that leverages light-activated photosensitizers to induce the release of endosome-entrapped molecules into the cytoplasm. By using CPP-photosensitizer (CPP-PS) conjugates, PCI effectively enhances the delivery of macromolecules like nucleic acids and proteins, allowing for rapid cytosolic entry while minimizing damage to healthy cells. This approach holds significant promise for advancing therapeutic strategies, particularly in oncology (Soe et al. 2020). Previously, endosomal escape was thought to be the primary route for CPPs to access the cytosol. When the plasma membrane is depolarized (which inhibits direct translocation), several CPPs show little to no cytosolic access, even though they are efficiently taken up via endocytosis (Serulla et al. 2023). This evidence implies that, in some cases, the direct translocation of CPPs across the plasma membrane may be a dominant pathway for cytosolic access, particularly for small, cationic CPPs such as TAT and arginine-rich peptides.

APPLICATIONS OF CPPs

CPPs have emerged as powerful tools in drug delivery, addressing many of the challenges traditionally faced in therapeutic applications. Their ability to traverse biological membranes, a significant barrier for many drugs, allows for the efficient delivery of therapeutic molecules such as proteins, nucleic acids, and small drugs directly into cells. By bypassing the need for conventional delivery methods, CPPs offer the potential to enhance the bioavailability and efficacy of treatments, particularly for diseases where intracellular targets are critical. Furthermore, their versatility in carrying diverse types of cargo, combined with their relatively low toxicity and ability to penetrate difficult biological barriers (e.g., the blood-brain barrier), positions CPPs as promising candidates for targeted drug delivery (Figure 6.2).

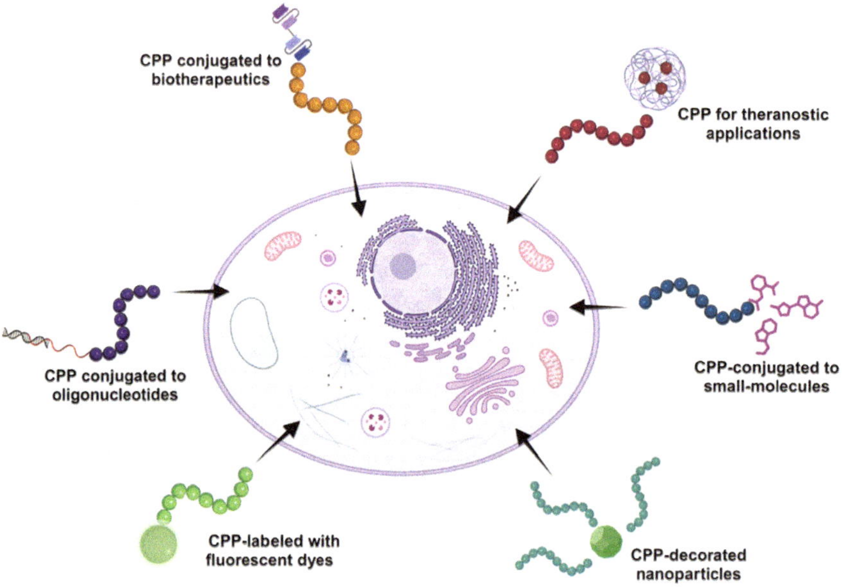

Figure 6.2 Current application of CPPs.

Enhancing drug penetration

CPPs play a crucial role in enhancing drug penetration into cancer cells, addressing one of the major challenges in cancer therapy. Many anticancer agents face obstacles such as poor membrane permeability and limited cellular uptake, which diminish their efficacy. CPPs address these issues by facilitating the transport of drugs across the plasma membrane of cancer cells. For instance, the TAT-peptide conjugated to doxorubicin (Dox) has been shown to significantly increase cellular uptake and cytotoxicity in breast cancer cell lines. In this study, the TAT-Dox conjugate was more potent (more effective at killing cancer cells) than the free drug at higher concentrations (Sen-caglar et al. 2015). TAT is also used to enhance the delivery of different therapeutic agents into cancer cells, including AlPcS (a photosensitizer for photodynamic therapy), and quantum dots (QDs, often used as drug carriers)—into cancer cells (Zhao et al. 2011).

Beyond crossing cellular barriers, CPPs also improve the distribution of drugs within tumors. In solid tumors, where drug diffusion is often restricted, F-G4-PTX-PEG-R8 (octa-arginine) conjugate has demonstrated enhanced tumor penetration and improved antitumor efficacy, as seen in 3D spheroids of HeLa (Rompicharla et al. 2018). Additionally, some CPPs equipped with Nuclear Localization Sequences (NLS) can target the nucleus directly, enhancing the nuclear delivery of drugs. An example of this is a long peptide with SV40 NLS containing the NLS and additional flanking sequences from the SV40 large T-antigen was effective in directing large IgM molecules into the nucleus quickly, within 30 minutes (Yoneda et al. 1992).

CPPs also enable the delivery of macromolecules that typically cannot enter cells on their own, such as therapeutic proteins and nucleic acids. For instance, CPPs enhance the intracellular delivery of peptides that can block the interactions between p53 and MDM2/MDMX. p53 is a tumor suppressor protein that regulates cell cycle arrest and apoptosis. However, its activity is often inhibited by interactions with MDM2 and MDMX in cancer cells. Researchers conjugated it to a CPP called cR10, which has a strong positive charge. CPPs facilitate the crossing of the cell membrane, helping the peptide enter cells (Philippe et al. 2024).

CPPs Enhance Cellular Uptake of siRNA and miRNA

In addition to protein delivery, CPPs have been instrumental in improving the cellular uptake of nucleic acid-based therapeutics. The main strategy for the CPP application for the delivery of nucleic acids into cells is a non-covalent strategy, the CPPs form a complex with the cargo due to electrostatic or hydrophobic interactions. Khaitov's group developed a therapeutic strategy to treat COVID-19 using siRNA, they created the siRNA called siR-7-EM (Khaitov et al. 2021). To enhance cellular uptake, they formulated siR-7-EM with a peptide dendrimer called KK-46. This combination helps siRNA enter cells more efficiently, particularly when administered via inhalation. The siR-7-EM/KK-46 complex was tested in Syrian hamsters and it showed that the inhaled siRNA formulation significantly reduced the viral load in the lungs. It also reduced lung inflammation, which is a key symptom of COVID-19, showing the treatment's potential to alleviate symptoms and limit damage (Khaitov et al. 2021).

CPPs can be Used to Enhance Nanoparticle-based Drug Delivery Systems

Nanoparticles have been widely used to improve intracellular drug concentration while avoiding toxicity in healthy tissue. Now, CPPs can be attached to nanoparticles to make nanoparticle-based drug delivery systems better. In recent years, many CPP-nanoparticle conjugates have been developed, highlighting their usefulness. A Magnetic Fluid Hyperthermia (MFH) method to treat prostate cancer cells with trimagnetic nanoparticles (TMNPs) was developed. Once TMNPs are inside prostate cancer cells, they produce heat that can kill cancer cells via hyperthermia. LN1 peptide conjugated to TMNPs and helps the TMNPs penetrate the prostate cancer cells, increasing the delivery efficiency. The study with this CPP-nanoparticle conjugate demonstrated that they can target cancer cells more specifically and deliver the hyperthermia effect with high precision and significantly induce apoptosis in prostate cancer cells (Nica et al. 2023).

Overcoming Drug Resistance in Cancer Cells

Overcoming drug resistance in cancer cells remains one of the biggest challenges in cancer treatment. CPPs can offer promising strategies to challenge this issue. One of the major mechanisms of drug resistance involving the cell membrane is decreased drug accumulation, which can result from various factors such as decreased drug influx, increased drug efflux, and membrane lipid alterations. CPPs can help drugs evade expulsion by multidrug resistance proteins (Peetla et al. 2013, Reygaert 2018). When drugs overcome these barriers, the drug concentration will also increase, enhancing the accumulation of drugs in cancer cells (Zhou et al. 2022). Moreover, CPPs can deliver inhibitors of resistance pathways based on targeting resistance mechanisms (Jin et al. 2022). In addition, CPPs facilitate the co-delivery of multiple agents to overcome resistance in combination therapy approaches (Zhou et al. 2022, Rowland et al. 2023).

Cancer cells develop mechanisms like efflux pumps to resist chemotherapy. These pumps actively remove therapeutic drugs from cells, reducing their effectiveness. To bypass efflux pumps by attaching a drug to a CPP, specifically D-octaarginine was developed. The conjugation of drugs to CPPs creates a single, water-soluble entity that can rapidly enter cells without being recognized by efflux pumps. The strategy has been successfully tested with paclitaxel (taxol) in both cellular and animal models with drug-resistant cancer. In ovarian cancer samples from patients, the taxol-octaarginine conjugates were effective at overcoming resistance and killing cancer cells (Vargas et al. 2014). Also, a CPP called [(WR)8WKβA] which is composed of alternating tryptophan (W), arginine (R) residues and additional lysine (K) residues, evaded recognition by these efflux pumps. It was attached to doxorubicin, and the [(WR)8WKβA]-Dox conjugate improved cancer cell-killing ability compared to free Dox. The conjugate inhibited cell viability of ovarian cancer by 59%, breast cancer MDA-MB-231 and MCF-7 by 71 and 77% at 5 μM, respectively. Dox alone inhibited cell viability of ovarian cancer by 35%, MDA-MB-231 and MCF-7 by 63 and 57%, respectively. Additionally, the conjugate also showed better activity against Dox-resistant cells (MES-SA/MX2), inhibiting 92% of the resistant cells' viability compared to only 15% for Dox alone at 5 μM. In general, this CPPs-drug conjugate enhanced the drug's cellular uptake, overcoming a common barrier in drug resistance (Zoghebi et al. 2022).

CPPs as Carriers of Imaging Agents

CPPs are proving to be valuable tools in enhancing diagnostic imaging, offering a range of applications that improve both the accuracy and effectiveness of cancer detection and monitoring. Compared to conventional imaging agents, CPPs offer several advantages, including enhanced cellular uptake, intracellular access, improved pharmacokinetics, better tumor retention, and the ability to image intracellular targets (Tripathi et al. 2018, Xie et al. 2020). Furthermore, CPPs enable more effective monitoring of treatment responses. CPPs have been successfully used to deliver contrast agents for various imaging diagnostic applications, such as conjugating with a dye allows surgeons to see and remove tumor tissues more effectively (Mander et al. 2022, Naffouje et al. 2022), carrying fluorescent dyes into tumor cells, enabling their visualization and detection (Mander et al. 2022, Yang et al. 2021, Zhang et al. 2019), delivering MRI contrast agents for enhanced tumor imaging (Li et al. 2019, Chen et al. 2019, Hingorani et al. 2020, Xie et al. 2022), or combining with radiotracers can be used for nuclear imaging of tumors (Knight et al. 2017, Veal et al. 2021, Goto et al. 2022). For example, by conjugating the tumor-targeting CPP p28 with an imaging agent indocyanine green (ICG), ICG-p28 can clearly localize the tumors and accurately identify tumor margins intraoperatively (Goto et al. 2022, Naffouje et al. 2022, Mander et al. 2022). Immunoconjugates including anti-γH2AX antibodies were modified with diethylenetriaminepentaacetic acid (DTPA) for labeling with the radioisotope indium-111 (111In) or the fluorophore Cy3 and TAT. The conjugate was incorporated to enhance nuclear translocation of the antibodies. Confocal microscopic images showed the localization of anti-γH2AX-TAT to γH2AX foci in irradiated breast cancer cells. The 111In-anti-γH2AX-TAT showed prolonged retention in cells compared to nonspecific antibody conjugates. In a mouse xenograft model of human breast cancer, the probes produced signals correlating with DNA damage after local irradiation or bleomycin treatment. Fluorescent and single-photon emission computed tomography signals were proportional to the radiation dose and γH2AX levels (Cornelissen et al. 2011, Knight et al. 2017). The study demonstrates the successful localization of these immunoconjugates to DNA damage sites *in vitro* and their ability to provide imaging signals *in vivo*, suggesting potential applications in cancer diagnosis and therapy monitoring.

FUTURE DIRECTIONS

Future directions for CPPs should focus on addressing their current limitations, such as enhancing cell selectivity, improving stability, and facilitating endosomal escape. Increasing the scope of clinical trials is also essential to evaluate the safety and efficacy of CPPs in human applications. Furthermore, integrating CPPs with nanotechnology, such as nanoparticles or liposomes, can potentially improve pharmacokinetics and enable the precise delivery of cargo molecules. Collectively, these strategies will pave the way for the more effective use of CPPs in cancer therapy and other clinical applications.

While there is no general strategy to improve the behavior of CPPs, this chapter discussed a variety of CPPs and their cargoes, providing valuable insights that enhance our understanding of these versatile molecules. This knowledge can inform future research directions and guide the rational design of next-generation CPPs, ultimately leading to improved efficacy in drug delivery and therapeutics.

ACKNOWLEDGMENTS

The illustrations were drawn by the BioRender. This work was supported by NIH/National Cancer Institute (R01CA272564 and R21CA280814 to TY).

REFERENCES

Agrawal, P., S. Bhalla, S.S. Usmani, S. Singh, K. Chaudhary, G.P.S. Raghava, et al. 2016. CPPsite 2.0: a repository of experimentally validated cell-penetrating peptides. Nucleic Acids Res. 44(D1): D1098–D1103.

Bennett, A.L., K.N. Cranford, A.L. Bates, C.R. Sabatini and H.-S. Lee. 2024. A molecular dynamics study of cell-penetrating peptide transportan-10 (TP10): Binding, folding and insertion to transmembrane state in zwitterionic membrane. Biochim. Biophys. Acta, Biomembr. 1866(1): 184218.

Botos, E., J. Klumperman, V. Oorschot, B. Ígyártó, A. Magyar, M. Oláh, et al. 2008. Caveolin-1 is transported to multi-vesicular bodies after albumin-induced endocytosis of caveolae in HepG2 cells. J. Cell. Mol. Med. 12(5a): 1632–1639.

Bottens, R.A. and T. Yamada. 2022. Cell-Penetrating Peptides (CPPs) as therapeutic and diagnostic agents for cancer. Cancers. 14(22): 5546.

Campeiro, J.D., W. Dam, G.G. Monte, L.C. Porta, L.C.G. de Oliveira, M.B. Nering, et al. 2019. Long term safety of targeted internalization of cell penetrating peptide crotamine into renal proximal tubular epithelial cells *in vivo*. Sci. Rep. 9(1): 3312.

Chehelgerdi, M., M Chehelgerdi, O.Q.B. Allela, R.D.C. Pecho, N. Jayasankar, D.P. Rao, et al. 2023. Progressing nanotechnology to improve targeted cancer treatment: overcoming hurdles in its clinical implementation. Mol. Cancer. 22(1): 169.

Chen, Y., Y. Fu, X. Li, H. Chen, Z. Wang and H. Zhang. 2019. Peptide-functionalized NaGdF4 nanoparticles for tumor-targeted magnetic resonance imaging and effective therapy. RSC Adv. 9(30): 17093–17100.

Chen, Y., L. Jia, G. Zhu, W. Wang, M. Geng, H. Lu, et al. 2022. Sortase A-mediated cyclization of novel polycyclic RGD peptides for $\alpha v\beta 3$ integrin targeting. Bioorg. Med. Chem. Lett. 73: 128888.

Conner, S.D. and S.L. Schmid. 2003. Regulated portals of entry into the cell. Nature. 422(6927): 37–44.

Cornelissen, B., V. Kersemans, S. Darbar, J. Thompson, K. Shah, K. Sleeth, et al. 2011. Imaging DNA damage *in vivo* using γH2AX-targeted immunoconjugates. Cancer Res. 71(13): 4539–4549.

Derakhshankhah, H. and S. Jafari. 2018. Cell penetrating peptides: A concise review with emphasis on biomedical applications. Biomed. Pharmacother. 108: 1090–1096.

Derossi, D., A.H. Joliot, G. Chassaing and A. Prochiantz. 1994. The third helix of the Antennapedia homeodomain translocates through biological membranes. J. Biol. Chem. 269(14): 10444–10450.

Deshayes, S., M.C. Morris, G. Divita and F. Heitz. 2006. Interactions of amphipathic CPPs with model membranes. Biochim. Biophys. Acta, Biomembr. 1758(3): 328–335.

Elliott, G. and P. O'Hare. 1997. Intercellular trafficking and protein delivery by a herpesvirus structural protein. Cell. 88(2): 223–233.

El-Sayed, A., S. Futaki and H. Harashima. 2009. Delivery of macromolecules using arginine-rich cell-penetrating peptides: Ways to overcome endosomal entrapment. AAPS J. 11(1): 13–22.

Erazo-Oliveras, A., N. Muthukrishnan, R. Baker, T.-Y. Wang and J.-P. Pellois. 2012. Improving the endosomal escape of cell-penetrating peptides and their cargos: Strategies and challenges. Pharmaceuticals. 5(11): 1177–1209.

Ezike, T.C., U.S. Okpala, U.L. Onoja, C.P. Nwike, E.C. Ezeako, O.J. Okpara, et al. 2023. Advances in drug delivery systems, challenges and future directions. Heliyon. 9(6): e17488.

Fernández-Carneado, J., M.J. Kogan, S. Pujals and E. Giralt. 2004. Amphipathic peptides and drug delivery. Pept. Sci. 76(2): 196–203.

Frankel, A.D. and C.O. Pabo. 1988. Cellular uptake of the tat protein from human immunodeficiency virus. Cell. 55(6): 1189–1193.

Fretz, M.M., N.A. Penning, S. Al-Taei, S. Futaki, T. Takeuchi, I. Nakase, et al. 2007. Temperature-, concentration- and cholesterol-dependent translocation of L- and D-octa-arginine across the plasma and nuclear membrane of CD_{34}+ leukaemia cells. Biochem. J. 403(Pt 2): 335–342.

Gao, C., S. Mao, H.J. Ditzel, L. Farnaes, P. Wirsching, R.A. Lerner, et al. 2002. A cell-penetrating peptide from a novel pVII–pIX phage-displayed random peptide library. Bioorg. Med. Chem. 10(12): 4057–4065.

Gautam, A., H. Singh, A. Tyagi, K. Chaudhary, R. Kumar, P. Kapoor, et al. 2012. CPPsite: A curated database of cell penetrating peptides. Database. 2012: bas015.

Gessner, I. and I. Neundorf. 2020. Nanoparticles modified with cell-penetrating peptides: Conjugation mechanisms, physicochemical properties, and application in cancer diagnosis and therapy. Int. J. Mol. Sci. 21(7): 2536.

Gkika, K.S., S. Noorani, N. Walsh and T.E. Keyes. 2021. Os(II)-bridged polyarginine conjugates: The additive effects of peptides in promoting or preventing permeation in cells and multicellular tumor spheroids. Inorg. Chem. 60(11): 8123–8134.

Gong, Z., S.P. Ikonomova and A.J. Karlsson. 2018. Secondary structure of cell-penetrating peptides during interaction with fungal cells. Protein Sci. 27(3): 702–713.

Gori, A., G. Lodigiani, S.G. Colombarolli, G. Bergamaschi and A. Vitali. 2023. Cell penetrating peptides: Classification, mechanisms, methods of study, and applications. ChemMedChem. 18(17): e202300236.

Goto, M., I. Ryoo, S. Naffouje, S. Mander, K. Christov, J. Wang, et al. 2022. Image-guided surgery with a new tumour-targeting probe improves the identification of positive margins. EBioMedicine. 76: 103850.

Green, M. and P.M. Loewenstein. 1988. Autonomous functional domains of chemically synthesized human immunodeficiency virus tat trans-activator protein. 55(6): 1179–1188.

Hao, M., L. Zhang and P. Chen. 2022. Membrane internalization mechanisms and design strategies of arginine-rich cell-penetrating peptides. Int. J. Mol. Sci. 23(16): 9038.

Herce, H.D. and A.E. Garcia. 2007. Molecular dynamics simulations suggest a mechanism for translocation of the HIV-1 TAT peptide across lipid membranes. Proc. Natl. Acad. Sci. 104(52): 20805–20810.

Herce, H.D., A.E. Garcia, J. Litt, R.S. Kane, P. Martin, N. Enrique, et al. 2009. Arginine-rich peptides destabilize the plasma membrane, consistent with a pore formation translocation mechanism of cell-penetrating peptides. Biophys. J. 97(7): 1917–1925.

Hingorani, D.V., F. Chapelin, E. Stares, S.R. Adams, H. Okada and E.T. Ahrens. 2020. Cell penetrating peptide functionalized perfluorocarbon nanoemulsions for targeted cell labeling and enhanced fluorine-19 MRI detection. Magn. Reson. Med. 83(3): 974–987.

Hirano, M., H. Yokoo, C. Goto, M. Oba, T. Misawa and Y. Demizu. 2023. Magainin 2-derived stapled peptides derived with the ability to deliver pDNA, mRNA, and siRNA into cells. Chem. Sci. 14(38): 10403–10410.

Holjencin, C.E., C.R. Feinberg, T. Hedrick, G. Halsey, R.D. Williams, P.V. Patel, et al. 2021. Advancing peptide siRNA-carrier designs through L/D-amino acid stereochemical modifications to enhance gene silencing. Mol. Ther. Nucleic Acids. 24: 462–476.

Hori, K., S. Higashida, T. Osaki, T. Kawano, H. Inaba, K. Matsuura, et al. 2022. Intracellular delivery and photothermal therapeutic effects of polyhistidine peptide-modified gold nanoparticles. J. Biotechnol. 354: 34–44.

Jin, P., J. Jiang, L. Zhou, Z. Huang, E.C. Nice, C. Huang, et al. 2022. Mitochondrial adaptation in cancer drug resistance: prevalence, mechanisms, and management. J. Hematol. Oncol. 15(1): 97.

Joliot, A., C. Pernelle, H. Deagostini-Bazin and A. Prochiantz. 1991. Antennapedia homeobox peptide regulates neural morphogenesis. Proc. Natl. Acad. Sci. 88(5): 1864–1868.

Kalafatovic, D. and E. Giralt. 2017. Cell-penetrating peptides: Design strategies beyond primary structure and amphipathicity. Mol.: J. Synth Chem. Nat. Prod. Chem. 22(11): 1929.

Khairkhah, N., A. Namvar and A. Bolhassani. 2023. Application of cell penetrating peptides as a promising drug carrier to combat viral infections. Mol. Biotechnol. 65(9): 1–16.

Khaitov, M., A. Nikonova, I. Shilovskiy, K. Kozhikhova, I. Kofiadi, L. Vishnyakova, et al. 2021. Silencing of SARS-CoV-2 with modified siRNA-peptide dendrimer formulation. Allergy. 76(9): 2840–2854.

Knight, J.C., S. Koustoulidou and B. Cornelissen. 2017. Imaging the DNA damage response with PET and SPECT. Eur. J. Nucl. Med. Mol. Imaging. 44(6): 1065–1078.

Komin, A., L.M. Russell, K.A. Hristova and P.C. Searson. 2017. Peptide-based strategies for enhanced cell uptake, transcellular transport, and circulation: Mechanisms and challenges. Adv. Drug Delivery Rev. 110–111: 52–64.

Langel, Ü. 2021. Cell-Penetrating Peptides and Transportan. Pharmaceutics. 13(7): 987.

Langel, Ü., M. Pooga, C. Kairane, M. Zilmer and T. Bartfai. 1996. A galanin-mastoparan chimeric peptide activates the Na^+, K^+-ATPase and reverses its inhibition by ouabain. Regul. Pept. 62(1): 47–52.

Lee, H., S.I. Lim, S.-H. Shin, Y. Lim, J.W. Koh and S. Yang. 2019a. Conjugation of cell-penetrating peptides to antimicrobial peptides enhances antibacterial activity. ACS Omega. 4(13): 15694–15701.

Lee, H.J., Y.-W. Huang, S.-H. Chiou and R.S. Aronstam. 2019b. Polyhistidine facilitates direct membrane translocation of cell-penetrating peptides into cells. Sci. Rep. 9(1): 9398.

Li, W., F. Nicol and F.C. Szoka. 2004. GALA: a designed synthetic pH-responsive amphipathic peptide with applications in drug and gene delivery. Adv. Drug Delivery Rev. 56(7): 967–985.

Li, Y., L. Hao, F. Liu, L. Yin, S. Yan, H. Zhao, et al. 2019. Cell penetrating peptide-modified nanoparticles for tumor targeted imaging and synergistic effect of sonodynamic/HIFU therapy. Int. J. Nanomed. 14: 5875–5894.

Liu, Y., Z. Lu, L. Mei, Q. Yu, X. Tai, Y. Wang, et al. 2017. Tandem peptide based on structural modification of poly-arginine for enhancing tumor targeting efficiency and therapeutic effect. ACS Appl. Mater. Interfaces. 9(3): 2083–2092.

Lo, S.L., and S. Wang. 2008. An endosomolytic Tat peptide produced by incorporation of histidine and cysteine residues as a nonviral vector for DNA transfection. Biomaterials. 29(15): 2408–2414.

Longoria-García, S., C.N. Sánchez-Domínguez and H.L. Gallardo-Blanco. 2022. Recent applications of cell-penetrating peptide guidance of nanosystems in breast and prostate cancer. Oncol. Lett. 23(3): 103.

Lorents, A., P.K. Kodavali, N. Oskolkov, Ü. Langel, M. Hällbrink and M. Pooga. 2012. Cell-penetrating peptides split into two groups based on modulation of intracellular calcium concentration. J. Biol. Chem. 287(20): 16880–16889.

Lucana, M.C., Y. Arruga, E. Petrachi, A. Roig, R. Lucchi and B. Oller-Salvia. 2021. Protease-resistant peptides for targeting and intracellular delivery of therapeutics. Pharmaceutics. 13(12): 2065.

Lunde, P.K., O. Manfra, T.P. Stole, M. Lunde, M. Martinsen, C.R. Carlson, et al. 2023. Polyarginine cell-penetrating peptides bind and inhibit SERCA2. Cells. 12(19): 2358.

Mander, S., S.A. Naffouje, J. Gao, W. Li, K. Christov, A. Green, et al. 2022. Tumor-targeting cell-penetrating peptide, p28, for glioblastoma imaging and therapy. Front. Oncol. 12: 940001.

Maniti, O., E. Blanchard, G. Trugnan, A. Lamazière and J. Ayala-Sanmartin. 2012. Metabolic energy-independent mechanism of internalization for the cell penetrating peptide penetratin. Int. J. Biochem. Cell Biol. 44(6): 869–875.

Mello, L.R. de, L. Porosk, T.C. Lourenço, B.B.M. Garcia, C.A.R. Costa, S.W. Han, et al. 2021. Amyloid-like self-assembly of a hydrophobic cell-penetrating peptide and its use as a carrier for nucleic acids. ACS Appl. Bio Mater. 4(8): 6404–6416.

Mohammadi, S., J.S. Mojarrad, P. Zakeri-Milani, A. Shirani, S. Mussa Farkhani, N. Samadi, et al. 2015. Synthesis and in vitro evaluation of amphiphilic peptides and their nanostructured conjugates. Adv. Pharm. Bull. 5(1): 41–49.

Morris, M.C., P. Vidal, L. Chaloin, F. Heitz and G. Divita. 1997. A new peptide vector for efficient delivery of oligonucleotides into mammalian cells. Nucleic Acids Res. 25(14): 2730–2736.

Mudhakir, D. and H. Harashima. 2009. Learning from the viral journey: How to enter cells and how to overcome intracellular barriers to reach the nucleus. AAPS J. 11(1): 65–77.

Naffouje, S., M. Goto, I. Ryoo, A. Green, T.K.D. Gupta and T. Yamada. 2022. A method of tumor *in vivo* imaging with a new peptide-based fluorescent probe. Methods Mol. Biol. 2394: 857–865.

Niazi, M., P. Zakeri-Milani, M. Soleymani-Goloujeh, A. Mohammadi, M. Sarfraz, R. Löbenberg, et al. 2021. Effects of self-assembled cell-penetrating peptides and their nano-complexes on ABCB1 expression and activity. Iran. J. Basic Med. Sci. 24(3): 383–390.

Nica, V., A. Marino, C. Pucci, Ö. Şen, M. Emanet, D. De Pasquale, et al. 2023. Cell-membrane-coated and cell-penetrating peptide-conjugated trimagnetic nanoparticles for targeted magnetic hyperthermia of prostate cancer cells. ACS Appl. Mater. Interfaces. 15(25): 30008–30028.

Oehlke, J., A. Scheller, B. Wiesner, E. Krause, M. Beyermann, E. Klauschenz, et al. 1998. Cellular uptake of an α-helical amphipathic model peptide with the potential to deliver polar compounds into the cell interior non-endocytically. Biochim. Biophys. Acta, Biomembr. 1414(1): 127–139.

Oh, D., A. Nasrolahi Shirazi, K. Northup, B. Sullivan, R.K. Tiwari, M. Bisoffi, et al. 2014. Enhanced cellular uptake of short polyarginine peptides through fatty acylation and cyclization. Mol. Pharmaceutics 11(8): 2845–2854.

Ohgita, T., T. Takechi-Haraya, R. Nadai, M. Kotani, Y. Tamura, K. Nishikiori, et al. 2019. A novel amphipathic cell-penetrating peptide based on the N-terminal glycosaminoglycan binding region of human apolipoprotein E. Biochim. Biophys. Acta - Biomembr. 1861(3): 541–549.

Ouyang, J., Y. Sheng and W. Wang. 2022. Recent advances of studies on cell-penetrating peptides based on molecular dynamics simulations. Cells. 11(24): 4016.

Peetla, C., S. Vijayaraghavalu and V. Labhasetwar. 2013. Biophysics of cell membrane lipids in cancer drug resistance: Implications for drug transport and drug delivery with nanoparticles. Adv. Drug Delivery Rev. 65(13): 1686–1698.

Peng, Y.-Y., H. Hu, D. Diaz-Dussan, J. Zhao, X. Hao and R. Narain. 2022. Glycopolymer–cell-penetrating peptide (cpp) conjugates for efficient epidermal growth factor receptor (EGFR) silencing. ACS Macro Lett. 11(4): 580–587.

Peters, P.J., A. Mironov, D. Peretz, E. van Donselaar, E. Leclerc, S. Erpel, et al. 2003. Trafficking of prion proteins through a caveolae-mediated endosomal pathway. J. Cell Biol. 162(4): 703–717.

Philippe, G.J.-B., Y.-H. Huang, A. Mittermeier, C.J. Brown, Q. Kaas, S.R. Ramlan, et al. 2024. Delivery to, and reactivation of, the p53 pathway in cancer cells using a grafted cyclotide conjugated with a cell-penetrating peptide. J. Med. Chem. 67(2): 1197–1208.

Poillot, C., H. Bichraoui, C. Tisseyre, E. Bahemberae, N. Andreotti, J.-M. Sabatier, et al. 2012. Small efficient cell-penetrating peptides derived from scorpion toxin maurocalcine. J. Biol. Chem. 287(21): 17331–17342.

Pooga, M., M. Hällbrink, M. Zorko and U. Langel. 1998. Cell penetration by transportan. FASEB J. 12(1): 67–77.

Preto, A.J., A.B. Caniceiro, F. Duarte, H. Fernandes, L. Ferreira, J. Mourão, et al. 2024. POSEIDON: Peptidic objects sequence-based interaction with cellular DOmaiNs: A new database and predictor. J. Cheminf. 16(1): 18.

Qi, X., T. Droste and C.C. Kao. 2011. Cell-penetrating peptides derived from viral capsid proteins. Mol. Plant-Microbe Interact. 24(1): 25–36.

Rádis-Baptista, G. 2021. Cell-penetrating peptides derived from animal venoms and toxins. Toxins. 13(2): 147.

Ramaker, K., M. Henkel, T. Krause, N. Röckendorf and A. Frey. 2018. Cell penetrating peptides: a comparative transport analysis for 474 sequence motifs. Drug Deliv. 25(1): 928–937.

Regberg, J., A. Srimanee, M. Erlandsson, R. Sillard, D.A. Dobchev, M. Karelson, et al. 2014. Rational design of a series of novel amphipathic cell-penetrating peptides. Int. J. Pharm. 464(1): 111–116.

Reygaert, W.C. 2018. An overview of the antimicrobial resistance mechanisms of bacteria. Adv. Drug. Deliv. Rev. 4(3): 482–501.

Rhee, M. and P. Davis. 2006. Mechanism of uptake of C105Y, a novel cell-penetrating peptide. J. Biol. Chem. 281(2): 1233–1240.

Richard, J.P., K. Melikov, E. Vives, C. Ramos, B. Verbeure, M.J. Gait, et al. 2003. Cell-penetrating Peptides: A REEVALUATION OF THE MECHANISM OF CELLULAR UPTAKE*. J. Biol. Chem. 278(1): 585–590.

Rompicharla, S.V.K., P. Kumari, B. Ghosh and S. Biswas. 2018. Octa-arginine modified poly(amidoamine) dendrimers for improved delivery and cytotoxic effect of paclitaxel in cancer. Artif. Cells Nanomed. Biotechnol. 46(sup2): 847–859.

Rowland, L., H.-B. Marjault, O. Karmi, D. Grant, L.J. Webb, A. Friedler, et al. 2023. A combination of a cell penetrating peptide and a protein translation inhibitor kills metastatic breast cancer cells. Cell Death Discov. 9(1): 1–11.

Ruseska, I. and A. Zimmer. 2020. Internalization mechanisms of cell-penetrating peptides. Beilstein J. Nanotechnol. 11: 101–123.

Rydström, A., S. Deshayes, K. Konate, L. Crombez, K. Padari, H. Boukhaddaoui, et al. 2011. Direct translocation as major cellular uptake for CADY self-assembling peptide-based nanoparticles. PLoS ONE. 6(10): e25924.

Saar, K., M. Lindgren, M. Hansen, E. Eiríksdóttir, Y. Jiang, K. Rosenthal-Aizman, et al. 2005. Cell-penetrating peptides: A comparative membrane toxicity study. Anal. Biochem. 345(1): 55–65.

Sadeghian, I., R. Heidari, S. Sadeghian, M.J. Raee and M. Negahdaripour. 2022. Potential of cell-penetrating peptides (CPPs) in delivery of antiviral therapeutics and vaccines. Eur. J. Pharm. Sci. 169: 106094.

Sahni, A., Z. Qian and D. Pei. 2020. Cell-penetrating peptides escape the endosome by inducing vesicle budding and collapse. ACS Chem. Biol. 15(9): 2485–2492.

Samec, T., K.L. Alatise, J. Boulos, S. Gilmore, A. Hazelton, C. Coffin, et al. 2022. Fusogenic peptide delivery of bioactive siRNAs targeting CSNK2A1 for treatment of ovarian cancer. Mol. Ther. Nucleic Acids. 30: 95–111.

Scheller, A., J. Oehlke, B. Wiesner, M. Dathe, E. Krause, M. Beyermann, et al. 1999. Structural requirements for cellular uptake of α-helical amphipathic peptides. J. Pept. Sci. 5(4): 185–194.

Sen-caglar, G.P., U. Gunduz and S. Yalcin. 2015. Synthesis and cytoxicity of a TAT peptide-doxorubicin conjugate for breast cancer treatment. J. Drug Des. Res. 2(1): 1–7.

Ser, J., J.Y. Lee, Y.H. Kim and H. Cho. 2020. Enhanced efficacy of photodynamic therapy by coupling a cell-penetrating peptide with methylene blue. Int. J. Nanomed. 15: 5803–5811.

Serulla, M., P. Anees, A. Hallaj, E. Trofimenko, T. Kalia, Y. Krishnan, et al. 2023. Plasma membrane depolarization reveals endosomal escape incapacity of cell-penetrating peptides. Eur. J. Pharm. Biopharm. 184: 116–124.

Silva, S., K. Kurrikoff, Ü. Langel, A.J. Almeida and N. Vale. 2022. A Second Life for MAP, a Model Amphipathic Peptide. Int. J. Mol. Sci. 23(15): 8322.

Singh, T., D.H. Kang, T.W. Kim, H.J. Kong, J.S. Ryu, S. Jeon, et al. 2021. Intracellular delivery of oxaliplatin conjugate *via* cell penetrating peptide for the treatment of colorectal carcinoma *in vitro* and *in vivo*. Int. J. Pharm. 606: 120904.

Singh, T., A.S.N. Murthy, H.-J. Yang and J. Im. 2018. Versatility of cell-penetrating peptides for intracellular delivery of siRNA. Drug Deliv. 25(1): 2005–2015.

Smidt, J.M., L. Lykke, C.E. Stidsen, N. Pristovšek and K.V. Gothelf. 2024. Synthesis of peptide–siRNA conjugates via internal sulfonylphosphoramidate modifications and evaluation of their *in vitro* activity. Nucleic Acids Res. 52(1): 49–58.

Soe, T.H., K. Watanabe and T. Ohtsuki. 2020. Photoinduced endosomal escape mechanism: A view from photochemical internalization mediated by CPP-photosensitizer conjugates. Molecules. 26(1): 36.

Spicer, C.D., C. Jumeaux, B. Gupta and M.M. Stevens. 2018. Peptide and protein nanoparticle conjugates: versatile platforms for biomedical applications. Chem. Soc. Rev. 47(10): 3574–3620.

Takeuchi, T. and S. Futaki. 2016. Current understanding of direct translocation of arginine-rich cell-penetrating peptides and its internalization mechanisms. Chem. Pharm. Bull. 64(10): 1431–1437.

Tanaka, M., A. Kato, Y. Satoh, T. Ide, K. Sagou, K. Kimura, et al. 2012. Herpes simplex virus 1 VP22 regulates translocation of multiple viral and cellular proteins and promotes neurovirulence. J. Virol. 86(9): 5264–5277.

Tripathi, P.P., H. Arami, I. Banga, J. Gupta and S. Gandhi. 2018. Cell penetrating peptides in preclinical and clinical cancer diagnosis and therapy. Oncotarget. 9(98): 37252–37267.

Tsuchiya, K., K. Horikoshi, M. Fujita, M. Hirano, M. Miyamoto, H. Yokoo, et al. 2023. Development of hydrophobic cell-penetrating stapled peptides as drug carriers. Int. J. Mol. Sci. 24(14): 11768.

Vargas, J.R., E.G. Stanzl, N.N.H. Teng and P.A. Wender. 2014. Cell-penetrating, guanidinium-rich molecular transporters for overcoming efflux-mediated multidrug resistance. Mol. Pharmaceutics. 11(8): 2553–2565.

Veal, M., G. Dias, V. Kersemans, D. Sneddon, S. Faulkner and B. Cornelissen. 2021. A model system to explore the detection limits of antibody-based immuno-SPECT imaging of exclusively intranuclear epitopes. J. Nucl. Med. 62(11): 1537–1544.

Verdurmen, W.P.R. and R. Brock. 2011. Biological responses towards cationic peptides and drug carriers. Trends Pharmacol Sci. 32(2): 116–124.

Vivès, E., P. Brodin and B. Lebleu. 1997. A truncated HIV-1 Tat protein basic domain rapidly translocates through the plasma membrane and accumulates in the cell nucleus. J. Biol. Chem. 272(25): 16010–16017.

Wang, Yu., J. Xuan, W. Zhao, Z. Ding, L. Zhang, R. Du, et al. 2020. Smart and selective cancer-killing peptides with cell penetrating sequence and dual-targeting mechanism. Colloids Surf., A: Physicochem. Eng. Aspects. 586: 124185.

Xia, M.-C., L. Cai, S. Zhang and X. Zhang. 2018. A cell-penetrating ratiometric probe for simultaneous measurement of lysosomal and cytosolic pH change. Talanta. 178: 355–361.

Xia, M.-C., L. Cai, Y. Yang, S. Zhang and X. Zhang. 2019. Tuning the pKa of carboxyfluorescein with arginine-rich cell-penetrating peptides for intracellular pH imaging. Anal. Chem. 91(14): 9168–9173.

Xie, J., Y. Bi, H. Zhang, S. Dong, L. Teng, R.J. Lee, et al. 2020. Cell-penetrating peptides in diagnosis and treatment of human diseases: From preclinical research to clinical application. Front. Pharmacol. 11: 697.

Xie, X., W. Liu, W. Zhu, G. Zhang, Y. Dai, J. Wu, et al. 2022. A cell penetrating peptide-modified magnetic/fluorescent probe for *in vivo* tracking of mesenchymal stem cells. J. Biomed. Mater. Res. Part A. 110(12): 1881–1891.

Yang, S.-T., E. Zaitseva, L.V. Chernomordik and K. Melikov. 2010. Cell-penetrating peptide induces leaky fusion of liposomes containing late endosome-specific anionic lipid. Biophys. J. 99(8): 2525–253.

Yang, Y., R. Li, S. Zhang and X. Zhang. 2021. A fluorescent nanoprobe based on cell-penetrating peptides and quantum dots for ratiometric monitoring of pH fluctuation in lysosomes. Talanta. 227: 122208.

Yao, J., Y. Ma, W. Zhang, L. Li, Y. Zhang, L. Zhang, et al. 2017. Design of new acid-activated cell-penetrating peptides for tumor drug delivery. PeerJ 5: e3429. Available at: https: //peerj.com/articles/3429 [Accessed October 21, 2024].

Yoneda, Y., T. Semba, Y. Kaneda, R.L. Noble, Y. Matsuoka, T. Kurihara, et al. 1992. A long synthetic peptide containing a nuclear localization signal and its flanking sequences of SV40 T-antigen directs the transport of IgM into the nucleus efficiently. Exp. Cell. Res. 201(2): 313–320.

Yue, G., C. Wang, B. Liu, M. Wu, Y. Huang, Y. Guo, et al. 2020. Liposomes co-delivery system of doxorubicin and astragaloside IV co-modified by folate ligand and octa-arginine polypeptide for anti-breast cancer. RSC Adv. 10(20): 11573–11581.

Zhang, M., M. Li, W. Zhang, Y. Han and Y.-H. Zhang. 2019. Simple and efficient delivery of cell-impermeable organic fluorescent probes into live cells for live-cell superresolution imaging. Light Sci. Appl. 8(1): 73.

Zhang, X., K. Oglęcka, S. Sandgren, M. Belting, E.K. Esbjörner, B. Nordén, et al. 2010. Dual functions of the human antimicrobial peptide LL-37—Target membrane perturbation and host cell cargo delivery. Biochim. Biophys. Acta, Biomembr. 1798(12): 2201–2208.

Zhao, J.-F., J.-Y. Chen, L. Mi, P.-N. Wang and Q. Peng. 2011. Enhancement of intracellular delivery of anti-cancer drugs by the tat peptide. Ultrastruct. Pathol. 35(3): 119–123.

Zhong, G., Z. Xu, R. Yang, S. Zhang, L. Li, M. Wu, et al. 2018. An arginine-rich cell penetrating peptide contained anti-gelatinase scFv-LDM fusion protein shows potent antitumor efficacy in pancreatic cancer. J. Cancer. 9(4): 674–682.

Zhou, J., W. Liu, R.-C. Pong, G. Hao, X. Sun and J.-T. Hsieh. 2010. Analysis of oligo-arginine cell-permeable peptides uptake by prostate cells. Amino Acids. 42(4): 1253–1260.

Zhou, M., X. Zou, K. Cheng, S. Zhong, Y. Su, T. Wu, et al. 2022. The role of cell-penetrating peptides in potential anti-cancer therapy. Clin. Transl. Med. 12(5): e822.

Zoghebi, K., H.M. Aliabadi, R.K. Tiwari and K. Parang. 2022. [$(WR)_8WK\beta A$]-Doxorubicin conjugate: A delivery system to overcome multi-drug resistance against doxorubicin. Cells. 11(2): 301.

Chapter 7

Recent Advances and Applications in Self-Assembling Short-Chain Elastin-Like Peptide Analogs

Takeru Nose* and Keitaro Suyama
Faculty of Arts and Science, Kyushu University, Fukuoka, 819-0395, Japan

INTRODUCTION

Elastin is a key protein abundantly found in the connective tissue of the body and is known for its remarkable elastic properties. Elastin is an insoluble protein complex that is difficult to solubilize, making research on its structure and function challenging and slow (Leach et al. 2005). Tropoelastin, the precursor protein and the main component of elastin (Debelle and Tamburro 1999), is synthesized by fibroblasts and smooth muscle cells along with fibrillin and forms complexes that organize it into functional elastic fibers (Vrhovski and Weiss 1998). Tropoelastin is primarily composed of Glycine (Gly, G), Valine (Val, V), Alanine (Ala, A), and Proline (Pro, P), which are rich in hydrophobic domains

*For Correspondence: nose@artsci.kyushu-u.ac.jp

of the protein that contribute to its elasticity. Moreover, its multiple cross-linking domains form cross-links via desmosine and isodesmosine through Lysine (Lys, K) residues (Partridge et al. 1966). These crosslinks create a resilient and flexible network that is crucial for the normal function of the arteries, lungs, skin, and other tissues that can get damaged in the body. However, elastin has a very low turnover rate and when damaged, these elastic tissues cannot fully regenerate, thereby contributing to aging and the loss of elasticity in tissues.

This review focuses on the Elastin-Like Peptides (ELPs) that are derived from repetitive sequence motifs observed in the hydrophobic domains that contribute to the unique characteristics of elastin (Vrhovski and Weiss 1998). The high Gly and Pro content in these repetitive sequences contribute to the flexibility and elasticity of the protein. Gly allows for tight packing of the peptide chains, whereas Pro induces bending and turning in the elastin protein chain. Research has focused on these repetitive sequences to elucidate the elastic properties of elastin and the self-aggregation characteristics of tropoelastin, which are vital for elastin fiber formation (Urry et al. 2010, Yeo et al. 2011). ELPs are derived from the internal amino acid sequences of tropoelastin and exhibit temperature-dependent reversible self-assembly. Various studies on ELPs have utilized these repetitive sequences as motifs, and extensive research is being conducted on the properties and structures of the amino acids that constitute these different ELP motifs, as well as on the functions and applications of ELPs.

BASICS ON ELASTIN-LIKE PEPTIDES

ELPs are synthetic polymers that mimic the amino acid sequence and properties of natural tropoelastins. They have been extensively studied owing to their unique characteristics and potential applications. Regardless of whether ELPs are chemically synthesized or prepared through expression systems, the most common studied feature is their temperature-responsive behavior and resulting aggregation properties, which are similar to those of tropoelastin (Yeo et al. 2011). This indicates that ELPs are capable of reversible self-assembly [referred to as "coacervation"] at a Lower Critical Solution Temperature (LCST) (Fernandez-Trillo et al. 2008). The self-assembly ability of ELPs is frequently evaluated using the transition temperature (T_t) (Urry 1992, 1997). Below T_t, ELPs are soluble in water, whereas above T_t, they become insoluble and form aggregates (Figure 7.1).

This characteristic of ELPs is useful for creating biomaterials that respond to changes in temperature.

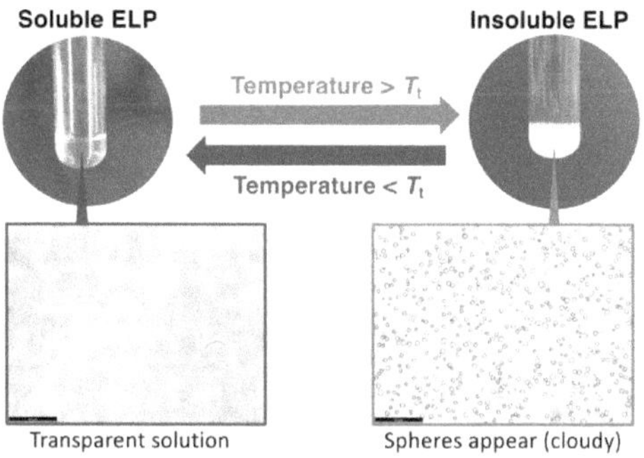

Figure 7.1 Temperature-responsive ELP forming aggregates.

ELPs are typically composed of repetitive hydrophobic amino acids sequences, such as Gly-Gly-Val-Pro (GGVP), Val-Pro-Gly-Val-Gly (VPGVG), and Gly-Val-Gly-Val-Ala-Pro (GVGVAP) which are present in tropoelastin (Vrhovski and Weiss 1998). These repetitive sequences are considered to be responsible for this self-assembly. Among these, the VPGVG repetitive sequence is widely used as the fundamental sequence of representative ELPs. The relationship between ELP sequences and their physicochemical properties has been extensively investigated using the pentapeptide Val-Pro-Gly-X-Gly $(VPGXG)_n$, where n is the number of repeats within the pentapeptide sequence, and X [called "guest residue"] can be any amino acid except Pro (Urry et al. 1992) (Figure 7.2).

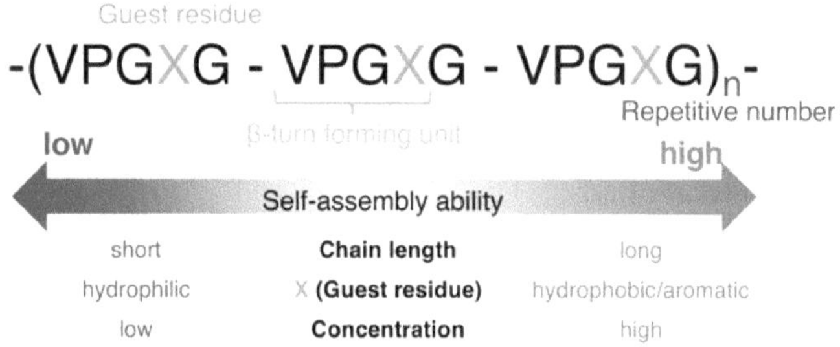

Figure 7.2 Various factors affecting the self-assembly activity of ELP.

The temperature-responsive behavior of ELPs is characterized by several intrinsic factors, such as the amino acid sequence (Urry and Pattanaik 1997, Keeley et al. 2002, Girotti et al. 2004) and the number of peptide repeats (Girotti et al. 2004, Meyer and Chilkoti 2004, McDaniel et al. 2013a); as well as extrinsic factors, such as the concentration of the peptide (Meyer and Chilkoti 2004, McDaniel et al. 2013a), pH of the solution (MacKay et al. 2010), and ionic strength (Luan et al. 1991). Therefore, the characteristics of ELPs can be modified by changing either X or n (Rodriguez-Cabello et al. 2018). In addition to their hydrophobicity, the crosslinking capabilities and pH responsiveness of

ELPs can be regulated by changing the guest residues (MacKay et al. 2010, Rodriguez-Cabello et al. 2018). Thus, it is possible to precisely control the physicochemical and biological properties of ELPs by changing their sequence composition to meet the requirements of biomaterials for various applications.

Based on research work by Urry (Urry 1992, 1997, Urry et al. 1992, 2010, Urry and Pattanaik 1997), it was determined that ELPs exhibit mechanical properties (including elasticity and resilience) and temperature responsive properties similar to those of natural elastin. Therefore, ELPs are considered suitable materials for applications in the field of tissue engineering, protein purification, and Drug Delivery Systems (DDS). Furthermore, ELPs can be readily combined with other materials, such as other proteins, synthetic polymers, and nanoparticles (NPs), to create hybrid materials with enhanced properties (Garanger and Lecommandoux 2022). In addition, ELPs derived from the amino acid sequence of human tropoelastin have low antigenicity, high biocompatibility, and biodegradability, making them safe for use in medical and biological applications (Urry et al. 1991b, Shah et al. 2012). Research on ELPs with these properties has shown that they are versatile and promising materials in a wide range of fields, including medicine, biotechnology, and materials science.

DEVELOPMENT OF SHORT-CHAIN ELASTIN PEPTIDES

Since the advent of recombinant DNA technology in the 1980s, ELPs have been prepared as genetically engineered proteins (McPherson et al. 1992, McDaniel et al. 2013a). This is because the expression of ELPs as recombinant proteins and peptides using *Escherichia coli* is one of the most cost-effective synthetic methods (Ibanez-Fonseca et al. 2019). Additionally, this review also extensively discusses ELPs synthesized through chemical methods, which allow for easy chemical modification of peptides, arbitrary alterations in amino acid sequences, and the introduction of unnatural amino acids.

Controlling the Self-assembling Properties of ELPs by Sequence Design

The reversible temperature-responsive self-aggregating properties of ELPs are derived from the repeated sequences of hydrophobic amino acids. However, this characteristic cannot be guaranteed simply by having a repeating sequence composed of hydrophobic amino acids. In most cases, ELPs composed of poly-$(VPGVG)_n$ sequences have been employed as long-chain peptides with $n \geq 40$ (200 amino acid residues) to achieve high coacervation ability (Urry 1997, Maeda et al. 2011). The process of self-assembly has been described as a phase separation between water and ELP molecules accompanied by a rise in temperature, driven by the difference in the entropy of the system in both states (Juanes-Gusano et al. 2022). In an aqueous solution below T_t, ELPs are fully hydrated as a result of hydrophobic hydration and are soluble. However, with temperatures above the T_t, ELPs release hydrating water molecules and undergo phase separation by forming hydrophobic interactions between the ELP molecules. Several theories have been proposed regarding the conformational changes involved in phase separation. Several studies have suggested that the ELPs change conformation from a random coil to a β-spiral structure (a helical structure that consists of consecutive β-turns) (Urry 1988, 1997, Li and Daggett 2003). In contrast, ELPs have been considered as a type of Intrinsically Disordered Protein (IDP) that exhibit phase separation without forming a single ordered 3D structure (Roberts et al. 2015. Ibanez-Fonseca et al. 2019, Juanes-Gusano et al. 2022). These studies do not claim that ELPs adopt a completely random structure upon self-assembly, instead ELPs are considered to exist as a wide range of nonrandom structures characterized by the presence of β-turns. In both theories (β-spiral and IDPs), type II β-turns that are formed by the VPGVG (especially VPGV, which includes tandem PG) sequence is considered important (Figure 7.3).

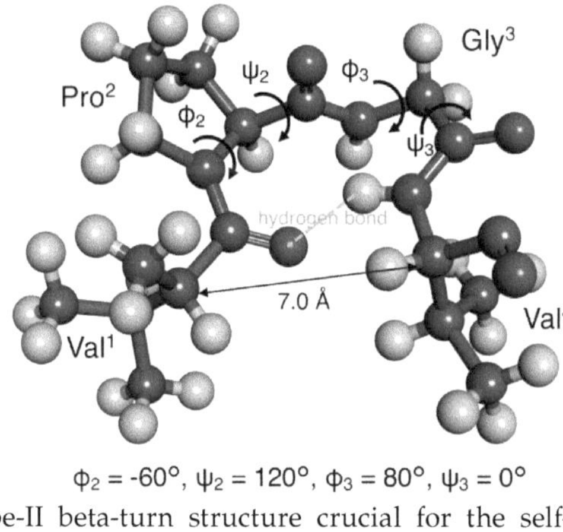

Figure 7.3 Type-II beta-turn structure crucial for the self-assembly of ELP centered around the Pro-Gly sequence.

The hydrogen bond between the carbonyl oxygen of the first residue and the amino hydrogen of the fourth residue stabilizes the turn structure. By altering the amino acid sequence of the guest residue, the secondary structure can either maintain its order or collapse.

Based on this, a previous study investigated the dependence of T_t on the sequence length and Ala content of guest residues using $(VPGXG)_n$ (X = A or V) (McDaniel et al. 2013a). Linear decreases in T_t were observed with the substitution of Ala residues for relatively hydrophobic Val residues, and nonlinear decreases were caused by sequence elongation. In another study using short synthetic ELPs containing X = Leucine (Leu, L), Isoleucine (Ile, I), and Phenylalanine (Phe, F), self-assembling properties were obtained even with short chain lengths using hydrophobic amino acids (Nuhn and Klok 2008). The dependence of the self-assembly properties on the amino acid sequence or content of ELP was also analyzed by measuring T_t of $C(VPGXG)_{10}$ (X = A, V, L, and I) (Aladini et al. 2016). In this study, the T_t of $C((VPGIG)_3(VPGLG)_2)_2$, which consisted of two repeating sequences, was lower than that of $C(VPGIG)_{10}$. Thus, the stereochemical structure of the side chain in Ile is thought to affect its self-assembly properties. This result indicates that the local hydration state of the residues that influences self-assembly differs depending on the bulkiness and hydrophobicity of the side chain, and this difference can change the macroscopic phase transition behavior of ELPs (Kurzbach et al. 2013). Based on these investigations, sequence design has become a popular approach for obtaining favorable self-assembly properties of ELPs.

In this context, it was discovered that replacing the first V residue in each repeat with other amino acids could result in ELPs with shorter peptide chain lengths that exhibit temperature-responsive reversible aggregation. $(IPGVG)_n$ (n = 7 – 10) containing Ile instead of Val, showed significantly higher aggregation ability (Maeda et al. 2013). This was attributed to the higher hydrophobicity of Ile compared to Val. Moreover, substituting Ile with Phe further enhances aggregation, leading to temperature-responsive aggregating peptides with a chain length of 25 amino acid residues (Maeda et al. 2015). This enhancement of self-assembly ability could not be explained solely by the hydrophobicity of the amino acids, as Phe is less hydrophobic than Val according to the hydrophobicity indices (Kyte and Doolittle 1982). Therefore, in addition to hydrophobicity, the aromaticity of amino acid residues is an important factor affecting aggregation potency. It has been reported that the incorporation of aromatic guest residues such as tryptophan (Trp, W), tyrosine (Tyr, Y), or Phe affects the LCST-like transition via induced chain stiffness, hydrophobic interactions, and turn structures (Prhashanna et al. 2019, Taylor et al. 2020). Therefore, it is hypothesized that stacking interactions between nearby aromatic side-chain groups would facilitate the aggregation of ELP chains. To examine the effects of aromatic amino acid residues on self-assembly, a Phe residue in $(FPGVG)_5$ was substituted with various amino acids (Suyama et al. 2022). In this study, it was determined that aromatic amino acids larger than Phe, such as Tyr, Trp, when substituted Phe, enhanced the self-assembly ability. In contrast, replacing one Phe in $(FPGVG)_5$ with a non-aromatic amino acid significantly reduced its ability to self-assemble. These studies suggest that π-π interactions due to the aromatic group of the Phe, Tyr, and Trp residues may enhance coacervation activity.

In addition to studies on changing amino acid residues in the pentapeptide sequence, modifications have been performed to improve the hydrophobicity of ELPs by truncating relatively hydrophilic amino acid residues from repetitive units (Sumiyoshi et al. 2022b). The coacervation ability of $(FPGV)_n$, the analogs in which the terminal Gly residue was deleted from the (FPGVG) repeat sequence, was significantly stronger than that of $(FPGVG)_n$ with the same number of repeats. This strategy is effective for obtaining self-assembled ELPs with shorter chain lengths. However, the shortened analog $(FPG)_4$ formed irreversible aggregates without an increase in temperature. This result indicated that Phe-Pro-Gly-Val (or Xxx_1-Pro-Gly-Xxx_2) components, which are required to form the β-turn structure, are important for both the strength and reversibility of the temperature-responsive self-assembly ability.

Control of Self-assembling Property by Multimerization

One method for arbitrarily controlling the temperature-dependent self-assembly properties of ELPs through artificial molecular structure design is multimerization. In this strategy, ELPs are not synthesized as long linear peptides using an expression system, but are instead obtained by connecting multiple short ELPs through interpeptide bonds or via scaffold molecules. For example, the dimerization of ELPs via disulfide bonds between additional cysteine residues (Aladini et al. 2016), chemically induced dimerization of receptor proteins and their ligands (Dhandhukia et al. 2013), and trimerization via the fibrin trimer domain (foldon) of the T4 phage (Ghoorchian et al. 2010, Ghoorchain and Holland 2011) have been reported. Dendrimer formation through peptide bonds branched via Lys residues is another important technique for obtaining multimeric ELPs (Navon et al. 2016, Zhou et al. 2017). Such oligomerization is not only used to enhance aggregation properties owing to the simple increase in molecular weight, but also to add the properties of ELPs to other proteins or peptides.

The multimerization of $(FPGVG)_n$ to improve the self-assembly ability of short ELP analogs was investigated. It was observed that dimerized $(FPGVG)_5$ or $(WPGVG)_3$ analogs synthesized by forming a disulfide bond via the introduction of cysteine (Suyama et al. 2016, Tatsubo et al.2018) showed lower T_t values than that of the corresponding monomer (XPGVG) (X = F or W). Subsequently, other multimerized ELPs, including dimeric, trimeric, and tetrameric peptides, were prepared using nitrilotriacetic acid (NTA) or ethylenediaminetetraacetic acid (EDTA) as scaffolds (Suyama et al. 2021, Tanaka et al. 2023). Multimeric ELPs conjugated with NTA showed strong self-assembly abilities at quite a low peptide concentration. This strategy represents a form of convergent synthesis that can efficiently produce temperature-responsive molecules using short-chain ELPs that can be chemically synthesized. Additionally, during the study of EDTA-mediated ELP analogs, it was determined that FPGVG exhibited a self-assembly ability with only one repetition. This is an important insight, indicating that FPGV(G) sequences do not need to be long or linearly connected to exhibit self-assembly.

Controlling the Self-assembling Property of ELPs by Cyclization

The cyclization of peptides has been employed as a useful strategy to control the structure of peptide molecules and peptide-based supramolecular nanostructures (Horne et al. 2003, Eskandari et al. 2017). In addition, cyclic peptides offer several advantages, including higher stability towards proteolysis, avoidance of endosomal uptake, improved cell permeability, and enhanced receptor-binding affinity compared to those of their linear counterparts (Shirazi et al. 2013). Several studies on ELPs involving cyclized peptides have been reported. For example, Urry et al. conducted structural analyses of cyclized elastin-derived short peptides, such as cyclo-$(VPGVG)_n$ (n = 1–6, mainly 3), using Nuclear Magnetic Resonance (NMR) spectroscopy and X-ray crystal structure analysis (Urry et al. 1981, 1989). These studies revealed that the conformation of cyclic analogs of $(VPGVG)_n$ was quite similar to the corresponding linear peptides and was classified as a type-II β-turn when the repetition number (n) ranged from 3 to 6. The LCST behavior of cyclic ELP analogs was reported by Wang et al. using a cyclic ELP-brush polymer, which is an analog of cyclic $(VPGGGVPGIG)_2$ polymerized by ring-opening metathesis polymerization via a norbornene unit (Wang et al. 2015). It was hypothesized that the polymerized cyclic ELP would show an enhanced self-assembly ability compared to that of its linear version, as cyclic peptides are inherently more hydrophobic (lacking terminal amines and carboxylic acids). As expected, the polymerized cyclic ELP exhibited thermally responsive LCST behavior, whereas the corresponding linear analog did not exhibit LCST under the same conditions. In response to these findings, the self-assembly ability of cyclized $(FPGVG)_n$ (n = 1–5) was investigated, and the cyclization of short $(FPGVG)_n$ was determined to be an effective strategy for improving the self-assembly ability. For example, cyclic $(FPGVG)_5$ exhibited a significantly lower T_t and stronger self-assembly ability at low peptide concentrations compared to that of its linear analogs (Suyama et al. 2018). Additionally, cyclic $(FPGVG)_5$ showed a stronger self-assembly ability than acetyl-$(FPGVG)_5$-NH_2, which lacks ionizable functional groups. Therefore, these studies suggest that peptide cyclization enhances the self-assembly ability of ELP analogs compared with their linear counterparts.

Controlling the Self-assembling Property of ELPs by Artificial Modification

To control T_t of ELPs, the modification of the peptide structure with various molecules other than natural amino acids has been investigated. This strategy includes the modification of natural amino acids and introduction of other functional molecules or unnatural amino acids. Chemical modification methods for the side chains of amino acid residues have been developed to alter ELP hydrophobicity. In these studies, oxidation, or alkylation of methionine residues at the guest position was employed (Petitdemange et al. 2017a, b). In another example, two small molecules that exhibit Fluorescence Resonance Energy Transfer (FRET) were introduced into a short ELP (Pramounmat et al. 2022) to confirm the phase transition behavior. To enhance the self-assembly ability of short-chain ELPs, the *N*-terminal Phe of $(FPGVG)_5$ was replaced with several derivatives containing various substituents on the aromatic ring (Suyama et al. 2022). The self-assembly ability of $(FPGVG)_5$ increased as the size of the substituent on the aromatic ring of the *N*-terminal Phe residue increased, regardless of whether the substituent was electron-donating or electron-withdrawing. Therefore, the coacervation ability of short ELPs could be drastically altered by substitution with only one amino acid.

Chemical modification of amino acid residues is thus an effective technique for controlling ELP self-assembly. Especially, chemically synthesized short-chain ELPs may be suitable for chemical modification with functional molecules because the position of the reactive functional group can be precisely controlled. In addition to controlling the ability of self-assembly, modifying ELPs with bioactive small molecules or medicinal oligopeptides has expanded the potential applications of ELPs in the medical field.

APPLICATIONS OF ELPs

Recently, ELPs have been recognized as attractive biomimetic polymers for biological and biomedical applications owing to their stimuli-responsive characteristics and ease of production (in terms of sequence and chain length reproducibility, production yield, and possible production scalability). Since ELPs are temperature-responsive polypeptides, they can be easily fused to a target protein which can then be purified using a non-chromatographic method (described later). Furthermore, research on ELP applications indicated that there are currently several examples of short-chain peptides synthesized chemically, in addition to high-molecular-weight polypeptides produced by expression systems. ELPs have been investigated for a wide variety of applications, such as (i) protein purification, (ii) hydrogel formation, (iii) drug delivery, and (iv) scavenging of hazardous materials (Figure 7.4).

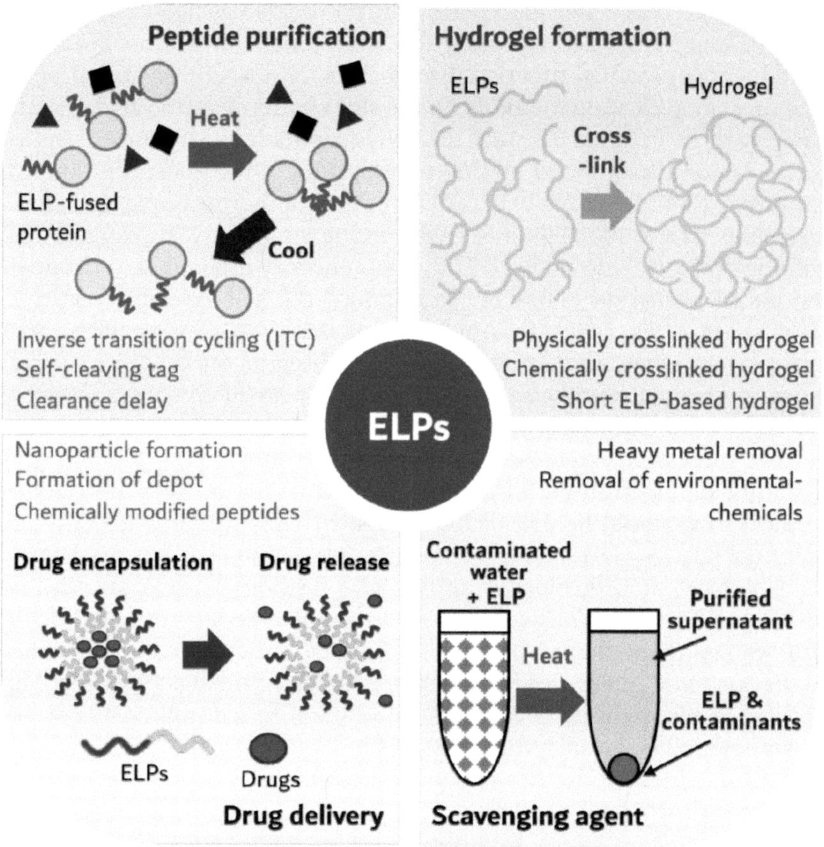

Figure 7.4 Various applications of ELPs.

The studies reviewed here are applications that utilize the temperature-responsiveness of ELP. Although not described in detail here, ELPs have also been investigated for various applications in tissue engineering (Nelson and Gilbert 2021) and artificial cells (Zhao et al. 2021).

Protein Purification

Recently, highly purified proteins have become necessary not only for the development of therapeutic drugs but also for several other applications in biotechnology and medical research. To easily and quickly obtain target proteins, overexpression in *E. coli* is a common technique for producing proteins. An efficient method for purifying recombinant proteins involves the use of ELPs as purification tags. ELPs retain their temperature-responsive reversible phase transition behavior even when fused to a protein. (Meyer and Chilkoti 1999, Meyer et al. 2001). Therefore, ELP-fused proteins can selectively precipitate from cellular contaminants above T_t. Subsequently, by dissolving precipitated recombinant proteins in water below their T_t, insoluble impurities can be removed. Purification of recombinant proteins by repetition of this process is termed Inverse Transition Cycling (ITC) (Hassouneh et al. 2010, Mullerpatan et al. 2020). Meyer and Chilkoti employed this method to enable direct precipitation of ELP-fused proteins (Meyer and Chilkoti 1999), which is regarded as an effective strategy in terms of purification performance compared to that of conventional methods (MacEwan and Chilkoti 2014, Gao et al. 2018). The temperature-responsive behavior of ELP-fused proteins can be adjusted by changing the length and amino acid composition of the ELP (Urry et al. 1991a, Meyer and Chilkoti 2004), the position to attach an ELP tag (*N*- or *C*-terminus of the target protein) (Christensen et al. 2009), and the presence of co-solutes (MacEwan and Chilkoti 2014). By adjusting these factors, ELPs can be used as purification tags for a wide range of proteins and peptides.

When using ELPs as protein purification tags, it may be necessary to adjust the ELP sequence before fusion or, in some cases, to later cleave the ELP tag from the target proteins to reduce its effect on the activity of the target proteins (Guo et al. 2023). In the early stages of ITC development, appropriate proteases were required to cleave the ELP tag from target proteins (Meyer and Chilkoti 1999). This can now also be achieved by incorporating a self-cleaving intein domain between the target protein and ELP tag. Wood et al. demonstrated that self-cleaving ELP-intein tags can provide highly purified target proteins by controlling ELP cleavage triggered by low pH and high temperature (Wood et al. 1999, Coolbaugh et al. 2017). Chilkoti and coworkers also reported recombinant fusion protein purification methods without chromatography, using a combination of ELP-fused proteins with intein or *Staphylococcus aureus* Sortase A (SrtA) transpeptidase (Ge et al. 2005, Bellucci et al. 2013). Briefly, the introduction of self-cleaving tags within the ELP system offers a reasonably low-cost, chromatography-free, and time-efficient method with potential applications for large-scale production of target proteins.

However, in some cases, the purified product which retains the ELP tag can increase the half-life and long-term storage stability of target protein (Floss et al. 2010). For example, it was demonstrated that an anti-Tumor Necrosis Factor (TNF)-α VHH (single-domain) antibody fused with ELP increased the hydrodynamic radius of the protein, delaying its clearance in the kidney and thereby increasing its half-life *in vivo* (Strohl 2015). It has also been reported that a binary tag composed of ELP and a histidine-tag is effective for the rapid purification and immobilization of β-glucosidase (Rong et al. 2018). Thus, whether the ELP tag should be cleaved after purification depends on its effect on the activity of the target protein and the function of the ELP after purification.

Formation of Hydrogels

Protein- and peptide-based hydrogels have great potential in tissue engineering owing to their ease of design and synthesis, diverse functionalization possibilities, and biomimetic and biocompatible properties (Li et al. 2019). In particular, hydrogels produced from synthetic ECM-mimetic materials (including ELPs) have attracted attention because they replicate the nanoscale architecture and material properties surrounding cells or neural tissue, facilitate tissue regeneration (Nelson and Gilbert 2021, Sharma et al. 2021), and show high biocompatibility and biodegradability compared to synthetic polymer-based hydrogels (Lee and Mooney 2001). The simplest method for forming architecture using ELPs is to raise the temperature of the ELP solution where an insoluble coacervate phase can be formed above the T_t as a result of the aggregation of ELP chains and the formation of an insoluble coacervate phase. However, ELP coacervates are often regarded as viscous fluids rather than hydrogels, because their structural stability and stiffness are limited in some applications (MacEwan and Chilkoti 2010). Therefore, ELP gelation is generally achieved using various crosslinking methods, including physical (Wu et al. 2005), chemical (Urry et al. 1991b, Vasconcelos et al. 2012), and light-induced crosslinking (Lee et al. 2001, Zhang et al. 2015).

The temperature-responsive properties of ELPs have attracted extensive interest in hydrogel fabrication. As a pioneering study in this field, Urry's group investigated the mechanical properties of two elastomeric polypentapeptides, $(GVGVP)_{251}$ and $(GVGIP)_{260}$, in equilibrium swelling states after being cross-linked by γ-irradiation (Lee et al. 2001). This study revealed that increased γ-radiation doses enhanced the modulus and toughness of the hydrogels and that temperature-induced phase transitions significantly impacted their mechanical properties, causing the hydrogels to become more viscoelastic above the T_t.

In recent years, elastin-based hydrogels have been developed for various applications, such as wound care, regenerative medicine, cardiac tissue engineering, ocular treatment, bone tissue regeneration, and neural tissue engineering (Vasconcelos et al. 2012, Sharma et al. 2021, Yang et al. 2024). In these studies, the self-assembly ability of ELP has been used for the formation and cross-linking of hydrogels. For example, hydrogel formation via physical crosslinking has been demonstrated using an ELP-based triblock copolymer (Wu et al. 2005). As another example, synthetic polymers combining ELPs and silk-like polymers (referred to as SELPs) have been exploited (Fernández-Colino et al. 2014). These polymers self-organize into a fibrous gel state from a sol state (sol-to-gel transition) with a temperature change. In addition, in some cases, chemically crosslinked ELP gels retained their temperature-responsive properties. McMillan and Conticello reported that hydrogels composed of Lys-containing ELPs and bis(sulfosuccinimidyl) suberate linkers showed temperature-triggered expansion or contraction (McMillan et al. 1999, McMillan and Conticello 2000). Temperature-responsive hydrogels have been actively investigated due to the sol–gel transition of ELP-based hydrogels in response to temperature, which makes them promising candidates for the development of stimuli-responsive or smart hydrogels.

Although ELP-based biopolymers and hydrogels have been extensively explored for biomedical applications, most of these studies have used long peptide sequences that were primarily synthesized using recombinant DNA technology. However, utilizing short ELPs as base materials for hydrogels as alternatives to native ECM is particularly interesting because it significantly reduces production and purification costs (Sharma et al. 2021). In addition, the self-assembling ability of short ELPs can be easily modified to develop biomaterials (Shi et al. 2011, Goktas et al. 2015). To date, only a few studies have reported the fabrication of hydrogels based on short-chain ELPs (Ding et al. 2011, Hassan et al. 2018, Cao et al. 2019). These studies could be of great help in developing novel biomaterials derived from elastin-based peptides, because short-sequence ELPs are easy to synthesize and possess improved tunable properties.

Drug Delivery

ELPs are regarded as promising candidates as drug carriers in the field of DDS owing to their stimulus-responsive characteristics and biocompatibility. To date, intensive research in materials science and therapeutics has led to the development of a wide range of stimuli-responsive drug carriers (Hong et al. 2020). In addition to their phase-transition ability, ELPs possess several advantages over other synthetic polymers. Specifically, because ELP sequences are composed of natural amino acids, they have low immunogenicity and platelet adhesion, are biodegradable, and can be easily obtained as recombinant or synthetic peptides. These peptides allow precise control over chain length and sequence for altering their properties (Chilkoti et al. 2006, Luginbuhl et al. 2017, Ibrahimova et al. 2021). Additionally, ELPs can be incorporated with various bio functional groups, including drugs, ligands, imaging agents, and cross-linking sites at any positions in their sequence (Christensen et al. 2013).

The typical DDS strategies using ELPs to date are to use (i) ELPs as drug carriers in the bloodstream and circulate them throughout the body, and (ii) three-dimensional networks composed of ELPs as a temporary storage area for drugs (frequently referred to as "depot") (Rodríguez-Cabello et al. 2016, Varanko et al. 2020). Considering the former, ELPs are used as soluble monomers or NPs. As described in the previous section, conjugation with ELP can enhance the stability of their fusion partners and delay the clearance of the fused molecules (Patel et al. 2007, Kaldis et al. 2013). The attachment of ELP increases the size of the therapeutic agent, thereby prolonging its half-life in serum and slowing renal filtration, consequently improving its biodistribution and pharmacokinetic profile compared to the free drug (Conrad et al. 2011). ELP block copolymers composed of hydrophobic and hydrophilic domains were developed to form NPs. This type of ELPs, which forms micelles with spherical or cylindrical morphologies, was first reported by Conticello and coworkers (Lee et al. 2000, Wright et al. 2002) and has been extensively studied as a potential drug carrier. ELP hybrid materials composed of ELPs fused with other agents can also be designed for assembly into NPs. Specifically, the conjugation of small hydrophobic molecules to ELPs can induce the formation of micelles that sequester small molecules in the core (known as attachment-directed assembly of micelles: ADAM) (McDaniel et al. 2013b). The utility of this technique has been demonstrated using ELPs fused with doxorubicin (Mastria et al. 2018). In contrast, depot-forming ELPs for DDS are designed to form insoluble coacervate-containing drugs after subcutaneous injection. Their phase-transition temperatures are adjusted between ambient and body temperatures by varying the polymer concentration, crosslinking conditions, and guest residues (Zhang et al. 2015). The ELP depot gradually dissolved from the surface and steadily released the drug into circulation. This strategy is particularly advantageous for treating diseases that require long-term treatment, such as cancer and chronic illnesses (Liu et al. 2012, Asai et al. 2012, Amiram et al. 2013, Luginbuhl et al. 2017).

ELPs have been investigated for various medical applications and drug delivery. However, short synthetic ELPs (less than ten pentapeptides) have rarely been utilized in temperature-responsive drug delivery systems because of their relatively high T_t (Bahniuk et al. 2018). To date, the application of short ELP derivatives has been investigated via conjugation with other hydrophobic molecules or functional polymers. Aluri et al. developed a series of peptide amphiphiles (PAs) by conjugating two saturated aliphatic acids with short ELPs $(VPGXG)_3$ (X = Ala, Val, Ile) that showed strong self-assembly ability (Aluri et al. 2012). These elastin-like peptide amphiphiles (ELPAs) formed temperature-responsive cylindrical micelles whose lengths could be tuned by the inclusion of other lipids and showed potential for the delivery of paclitaxel. ELP-oligonucleotide conjugates (ELP-ONs) have also been previously reported (Wang et al. 2021). ELP-ONs, composed of $(VPGFG)_n$ (n = 3 – 6) and a 32-mer oligonucleotide, form well-defined nanospheres that exhibit excellent reversible thermo-responsiveness, biocompatibility, encapsulation of drugs such as doxorubicin, and effective cellular uptake. This new class of short ELP conjugates is a potential platform for the development of nanostructures for targeted drug delivery.

Scavenging Agents for Hazardous Materials

Research on the application of ELPs as scavenging agents for harmful substances, primarily targeting heavy metals, has been conducted since the early 2000s. As the contamination of soil and groundwater with toxic heavy metal ions have become a significant environmental problem worldwide, diverse approaches to remove these contaminants from the environment have been investigated. Advances in genetic engineering have enabled the design of protein-based metal-binding biopolymers that can remove specific heavy metals by combining ELPs with metal-binding proteins. These polymers are highly desirable as metal removal agents because the adsorbed metals can be easily recovered as precipitates by inducing the aggregation of biopolymer-metal complexes through temperature changes.

To date, most studies on ELP-based metal removal agents have targeted cadmium. Several groups reported a synthetic polymer that combined ELP and polyhistidine clusters or synthetic phytochelatins (ECs) as the metal-binding domain, which could strongly bind to Cd^{2+} (Kostal et al. 2001, Lao et al. 2007, Chu et al. 2013). Their results indicated that the metal-binding ability of the synthetic polymer was due to the fused metal-binding domain and that ELP itself does not strongly bind to metal ions. Thus, it was suggested that the metal-binding capability of biopolymers can be tailored specifically for any target metal of interest simply by employing an appropriate metal-binding domain. Furthermore, this polymer could be reused by treatment with an acidic buffer solution to desorb and remove the bound cadmium ions. In another example, short-chain peptide analogs combining short-chain ELPs containing Phe (FPGVG or FPGV) and a cadmium-binding sequence (AADAAC) could remove cadmium ions from water with high efficiency (Sumiyoshi et al. 2022a, b). Furthermore, as scavenging agents for harmful substances other than cadmium, ELP analogs that bind mercury (Kostal et al. 2003) or endocrine-disrupting aromatic compounds (Suyama et al. 2023) have also been reported. Methods using ELPs would be beneficial for the development of environmentally friendly recovery agents for harmful substances.

CONCLUSION

ELPs derived from hydrophobic amino acid repeat sequences of tropoelastin have been explored for various applications. Currently, ELPs produced using expression systems are employed as research materials, yielding significant results, as described in this review. Meanwhile, more than half a century has passed since the development of the solid-phase peptide synthesis method, and now new chemical peptide synthesis techniques, characterized by their cost-effectiveness and high reaction efficiency, have been developed and established (Dawson et al. 1994, Okada et al. 2014, Agouridas et al. 2019, Muramatsu et al. 2019). These advancements have significantly facilitated the chemical synthesis of ELPs composed of hydrophobic amino acid repeat sequences that do not require side chain protection. Furthermore, progress has been made in developing short-chain ELPs with low molecular weights that exhibit temperature-responsive behavior and aggregative properties, leading to the expectation that in the near future, large-scale production of ELPs through chemical synthesis using amino acids as raw materials will become feasible, standing alongside biological expression systems. If realized, this would further advance the industrial application of ELPs.

ACKNOWLEDGMENTS

The authors would like to express their gratitude to Mr. Naoki Tanaka, M.Sc., for his valuable advice in writing this review.

REFERENCES

Agouridas, V., O.E. Mahdi, V. Diemer, M. Cargoët, J.-C.M. Monbaliu and O. Melnyk. 2019. Native chemical ligation and extended methods: Mechanisms, catalysis, scope, and limitations. Chem. Rev. 119: 7328–7443.

Aladini, F., C. Araman and C.F.W. Becker. 2016. Chemical synthesis and characterization of elastin-like polypeptides (ELPs) with variable guest residues. J. Pept. Sci. 22: 334–342.

Aluri, S., M.K. Pastuszka, A.S. Moses and J.A. MacKay. 2012. Elastin-like peptide amphiphiles form nanofibers with tunable length. Biomacromolecules. 13: 2645–2654.

Amiram, M., K.M. Luginbuhl, X. Li, M.N. Feinglos and Chilkoti A. 2013. A depot-forming glucagon-like peptide-1 fusion protein reduces blood glucose for five days with a single injection. J. Control. Release. 172: 144–151.

Asai, D., D. Xu, W. Liu, F.G. Quiroz, D.J. Callahan, M.R. Zalutsky, et al. 2012. Protein polymer hydrogels by *in situ*, rapid and reversible self-gelation. Biomaterials. 33: 5451–5458.

Bahniuk, M.S., A.K. Alshememry, S.V. Elgersma and L.D. Unsworth. 2018. Self-assembly/disassembly hysteresis of nanoparticles composed of marginally soluble, short elastin-like polypeptides. J. Nanobiotechnol. 16: 15.

Bellucci, J.J., M. Amiram, J. Bhattacharyya, D. McCafferty and Ashutosh Chilkoti. 2013. Three-in-one chromatography-free purification, tag removal, and site-specific modification of recombinant fusion proteins using sortase A and elastin-like polypeptides. Angew. Chem. Int. Ed. 52: 3703–3708.

Cao, M., Y. Shen, Y. Wang, X. Wang and D. Li. 2019. Self-Assembly of short elastin-like amphiphilic peptides: Effects of Temperature, molecular hydrophobicity and charge distribution. Molecules. 24: 202.

Chilkoti, A., T. Christensen and J.A. MacKay. 2006. Stimulus responsive elastin biopolymers: applications in medicine and biotechnology. Curr. Opin. Chem. Biol. 10: 652–657.

Christensen, T., M. Amiram, S. Dagher, K. Trabbic-Carlson, M.F. Shamji, L.A. Setton, et al. 2009. Fusion order controls expression level and activity of elastin-like polypeptide fusion proteins. Protein Sci. 18: 1377–1387.

Christensen, T., W. Hassouneh, K. Trabbic-Carlson and A. Chilkoti. 2013. Predicting transition temperatures of elastin-like polypeptide fusion proteins. Biomacromolecules. 14: 1514–1519.

Chu, H.S., J. Ryum and J.-I. Won. 2013. Cadmium detection by a thermally responsive elastin copolymer with metal-binding functionality. Enzyme Microb. Technol. 53: 189–193.

Conrad, U., I. Plagmann, S. Malchow, M. Sack, D.M. Floss, A.A. Kruglov, et al. 2011. ELPylated anti-human TNF therapeutic single-domain antibodies for prevention of lethal septic shock. Plant Biotechnol. J. 9: 22–31.

Coolbaugh, M.J., M.J. Shakalli Tang and D.W. Wood. 2017. High-throughput purification of recombinant proteins using self-cleaving intein tags. Anal. Biochem. 516: 65–74.

Dawson, P.E., T.W. Muir, I. Clark-Lewis and S.B.H. Kent. 1994. Synthesis of proteins by native chemical ligation. Science. 266: 776–779.

Debelle, L. and A.M. Tamburro. 1999. Elastin: Molecular description and function. Int. J. Biochem. Cell Biol. 31: 261–272.

Dhandhukia, J., I. Weitzhandler, W. Wang and J.A. MacKay. 2013. Switchable elastin-like polypeptides that respond to chemical inducers of dimerization. Biomacromolecules. 14: 976–985.

Ding, L., S. Wang, W. Wu, Y. Hu, C. Yang, M. Tan, et al. 2011. A supramolecular hydrogel inspired by elastin. Chin. J. Chem. 29: 2182–2186.

Eskandari, S., T. Guerin, I. Toth and R.J. Stephenson. 2017. Recent advances in self-assembled peptides: Implications for targeted drug delivery and vaccine engineering. Adv. Drug Delivery Rev. 110–111: 169–187.

Fernández-Colino, A., F.J. Arias, M. Alonso and J.C. Rodríguez-Cabello. 2014. Self-organized ECM-mimetic model based on an amphiphilic multiblock silk-elastin-like corecombinamer with a concomitant dual physical gelation process. Biomacromolecules. 15: 3781–3793.

Fernandez-Trillo, F., J.C. van Hest, J.C. Thies, T. Michon, R. Weberskirch and N.R. Cameron. 2008. Fine-tuning the transition temperature of a stimuli-responsive polymer by a simple blending procedure. Chem. Commun. 2230–2232.

Floss, D.M., K. Schallau, S. Rose-John, U. Conrad and J. Scheller. 2010. Elastin-like polypeptides revolutionize recombinant protein expression and their biomedical application. Trends Biotechnol. 28: 37–45.

Gao, H., X. Qi, D.J. Hart, S. Gao, H. Wang, S. Xu, et al. 2018. Three novel *Escherichia coli* vectors for convenient and efficient molecular biological manipulations. J. Agric. Food Chem. 66: 6123– 6131.

Garanger, E. and S. Lecommandoux. 2022. Emerging opportunities in bioconjugates of elastin-like polypeptides with synthetic or natural polymers. Adv. Drug Deliv. Rev. 191: 114589.

Ge, X., S.C. Yang, K. Trabbic-Carlson, B. Kim and A. Chilkoti. 2005. Filipe CDM. Self-cleavable stimulus responsive tags for protein purification without chromatography. J. Am. Chem. Soc. 127: 11228–11229.

Ghoorchian, A., J.T. Cole and N.B. Holland. 2010. Thermoreversible micelle formation using a three-armed star elastin-like polypeptide. Macromolecules. 43: 4340–4345.

Ghoorchian, A. and N.B. Holland. 2011. Molecular architecture influences the thermally induced aggregation behavior of elastin-like polypeptides. Biomacromolecules. 12: 4022–4029.

Girotti, A., J. Reguera, F.J. Arias, M. Alonso, A.M. Testera and J.C. Rodríguez-Cabello. 2004. Influence of the molecular weight on the inverse temperature transition of a model genetically engineered elastin-like pH-responsive polymer. Macromolecules. 37: 3396–3400.

Goktas, M., G. Cinar, I. Orujalipoor, S. Ide, A.B. Tekinay and M.O. Guler. 2015. Self-assembled peptide amphiphile nanofibers and peg composite hydrogels as tunable ECM mimetic microenvironment. Biomacromolecules. 16: 1247–1258.

Guo, Y., S. Liu, D. Jing, N. Liu and X. Luo. 2023. The construction of elastin-like polypeptides and their applications in drug delivery system and tissue repair. J. Nanobiotechnol. 21: 418.

Hassan, M., A.D. Martin and P. Thordarson. 2018. Engineering biocompatible scaffolds through the design of elastin-based short peptides. Chem. Plus Chem. 83: 47–52.

Hassouneh, W., T. Christensen and A. Chilkoti. 2010. Elastin-like polypeptides as a purification tag for recombinant proteins. Curr. Protoc. Protein Sci. 6: 6.11.1–6.11.16.

Hong, S., D.W. Choi, H.N. Kim, C.G. Park, W. Lee and H.H. Park. 2020. Protein-based nanoparticles as drug delivery systems. Pharmaceutics. 12: 604.

Horne, W.S., C.D. Stout and M.R. Ghadiri. 2003. A heterocyclic peptide nanotube. J. Am. Chem. Soc. 125: 9372–9376.

Ibanez-Fonseca, A., T. Flora, S. Acosta and J.C. Rodriguez-Cabello. 2019. Trends in the design and use of elastin-like recombinamers as biomaterials. Matrix Biol. 84: 111–126.

Ibrahimova, V., H. Zhao, E. Ibarboure, E. Garanger and S. Lecommandoux. 2021. Thermosensitive vesicles from chemically encoded lipid-grafted elastin-like polypeptides. Angew. Chem. Int. Ed. 60: 15036–15040.

Juanes-Gusano, D., M. Santos, V. Reboto, M. Alonso and J.C. Rodríguez-Cabello. 2022. Self-assembling systems comprising intrinsically disordered protein polymers like elastin-like recombinamers. J. Pept. Sci. 28: e3362.

Kaldis, A., A. Ahmad, A. Reid, B. McGarvey, J. Brandle, S. Ma, et al. 2013. High-level production of human interleukin-10 fusions in tobacco cell suspension cultures. Plant Biotechnol. J. 11: 535–545.

Keeley, F.W., C.M. Bellingham and K. Woodhouse. 2002. Elastin as a self-organizing biomaterial: Use of recombinantly expressed human elastin polypeptides as a model for investigations of structure and self-assembly of elastin. Philos. Trans. R. Soc. Lond. B. 357: 185–189.

Kostal, J., A. Mulchandani and W. Chen. 2001. Tunable biopolymers for heavy metal removal. Macromolecules. 34: 2257–2261.

Kostal, J., A. Mulchandani, E.G. Gropp and W. Chen. 2003. A temperature responsive biopolymer for mercury remediation. Environ. Sci. Technol. 37: 4457–4462.

Kurzbach, D., G. Platzer, T.C. Schwarz, M.A. Henen, R. Konrat, D. Hinderberger. 2013. Cooperative unfolding of compact conformations of the intrinsically disordered protein osteopontin. Biochemistry. 52: 5167–5175.

Kyte, J. and R.F. Doolittle. 1982. A simple method for displaying the hydropathic character of a protein. J. Mol. Biol. 157: 105–132.

Lao, U.L., A. Chen, M.R. Matsumoto, A. Mulchandari and W. Chen. 2007. Cadmium removal from contaminated soil by thermally responsive elastin (ELPEC20) biopolymers. Biotechnol. Bioeng. 98: 349–355.

Leach, J.B., J.B. Wolinsky, P.J. Stone and J.Y. Wong. 2005. Crosslinked α-elastin biomaterials: towards a processable elastin mimetic scaffold. Acta Biomater. 1: 155–164.

Lee, T.A.T., A. Cooper, R.P. Apkarian and V.P. Conticello. 2000. Thermo-reversible self-assembly of nanoparticles derived from elastin-mimetic polypeptides. Adv. Mater. 12: 1105–1110.

Lee, J., C.W. Macosko and D.W. Urry. 2001. Mechanical properties of cross-linked synthetic elastomeric polypentapeptides. Macromolecules. 34: 5968–5974.

Lee, K.Y. and D.J. Mooney. 2001. Hydrogels for tissue engineering. Chem. Rev. 101: 1869–1880.

Li, B. and V. Daggett. 2003. The molecular basis of the temperature- and pH-induced conformational transitions in elastin-based peptides. Biopolymers. 68: 121–129.

Li, J., R. Xing, S. Bai and X. Yan. 2019. Recent advances of self-assembling peptide-based hydrogels for biomedical applications. Soft Matter. 15: 1704–1715.

Liu, W., J. McDaniel, X. Li, D. Asai, F.G. Quiroz, J. Schaal, et al. 2012. Brachytherapy using injectable seeds that are self-assembled from genetically encoded polypeptides in situ. Cancer Res. 72: 5956–5965.

Luan, C.H., T.M. Parker, K.U. Prasad and D.W. Urry, D.W. 1991. Differential scanning calorimetry studies of NaCl effect on the inverse temperature transition of some elastin-based polytetra-, polypenta-, and polynonapeptides. Biopolymers. 31: 465–475.

Luginbuhl, K.M., J.L. Schaal, B. Umstead, E.M. Mastria, X. Li, S. Banskota, et al. 2017. One-week glucose control via zero-order release kinetics from an injectable depot of glucagon-like peptide-1 fused to a thermosensitive biopolymer. Nat. Biomed. Eng. 1: 0078.

MacEwan, S.R. and A. Chilkoti. 2010. Elastin-like polypeptides: Biomedical applications of tunable biopolymers. Biopolymers. 94: 60–77.

MacEwan, S.R. and A. Chilkoti. 2014. Applications of elastin-like polypeptides in drug delivery. J. Control Release. 28: 314–330.

MacKay, J.A., D.J. Callahan, K.N. FitzGerald and Chilkoti, A. 2010. Quantitative model of the phase behavior of recombinant pH-responsive elastin-like polypeptides. Biomacromolecules. 11: 2873–2879.

Maeda, I., Y. Fukumoto, T. Nose, Y. Shimohigashi, T. Nezu, Y. Terada, et al. 2011. Structural requirements essential for elastin coacervation: favorable spatial arrangements of valine ridges on the three-dimensional structure of elastin-derived polypeptide $(VPGVG)_n$. J. Pept. Sci. 17: 735–743.

Maeda, I., S. Taniguchi, J. Ebina, N. Watanabe, T. Hattori and T. Nose. 2013. Comparison between coacervation property and secondary structure of synthetic peptides, Ile-containing elastin-derived pentapeptide repeats. Protein Pept. Lett. 20: 905–910.

Maeda, I., S. Taniguchi N. Watanabe, A. Inoue Y. Yamasaki and T. Nose. 2015. Design of phenylalanine-containing elastin-derived peptides exhibiting highly potent self-assembling capability. Protein Pept. Lett. 22: 934–939.

Mastria, E.M., L.Y. Cai, M.J. Kan, X. Li, J.L. Schaal, S. Fiering, et al. 2018. Nanoparticle formulation improves doxorubicin efficacy by enhancing host antitumor immunity. J. Control. Release. 269: 364–373.

McDaniel, J.R., D.C. Radford and A. Chilkoti A. 2013a. A unified model for de novo design of elastin-like polypeptides with tunable inverse transition temperatures. Biomacromolecules. 14: 2866–2872.

McDaniel, J.R., J. Bhattacharyya, K.B. Vargo, W. Hassouneh, D.A. Hammer and A. Chilkoti. 2013b. Self-assembly of thermally responsive nanoparticles of a genetically encoded peptide polymer by drug conjugation. Angew. Chem. Int. Ed. 52: 1683–1687.

McMillan, R.A., K.L. Caran, R.P. Apkarian and V.P. Conticello. 1999. High-resolution topographic imaging of environmentally responsive, elastin-mimetic hydrogels. Macromolecules. 32: 9067–9070.

McMillan, R.A. and V.P. Conticello. 2000. Synthesis and characterization of elastin-mimetic protein gels derived from a well-defined polypeptide precursor. Macromolecules. 33(13): 4809–4821.

McPherson, D.T., C. Morrow, D.S. Minehan, J. Wu, E. Hunter and D.W. Urry. 1992. Production and purification of a recombinant elastomeric polypeptide, G-$(VPGVG)_{19}$-VPGV, from *Escherichia coli*. Biotechnol. Prog. 8: 347–352.

Meyer, D.E. and A. Chilkoti. 1999. Purification of recombinant proteins by fusion with thermally-responsive polypeptides. Nat. Biotechnol. 17: 1112–1115.

Meyer, D.E., K. Trabbic-Carlson and A. Chilkoti. 2001. Protein purification by fusion with an environmentally responsive elastin-like polypeptide: effect of polypeptide length on the purification of thioredoxin. Biotechnol. Prog. 17: 720–728.

Meyer, D.E. and A. Chilkoti. 2004. Quantification of the effects of chain length and concentration on the thermal behavior of elastin-like polypeptides. Biomacromolecules. 5: 846–851.

Mullerpatan, A., D. Chandra, E. Kane, P. Karande and S. Cramer. 2020. Purification of proteins using peptide-ELP based affinity precipitation. J. Biotechnol. 10: 59–67.

Muramatsu, W., T. Hattori and H. Yamamoto. 2019. Substrate-directed lewis-acid catalysis for peptide synthesis. J. Am. Chem. Soc. 141: 12288–12295.

Navon, Y., M. Zhou, J.B. Matson and R. Bitton. 2016. Dendritic elastin-like peptides: The effect of branching on thermoresponsiveness. Biomacromolecules. 17: 262–270.

Nelson, D.W. and Gilbert, R.J. 2021. Extracellular matrix-mimetic hydrogels for treating neural tissue injury: A focus on fibrin, hyaluronic acid, and elastin-like polypeptide hydrogels. Adv. Healthc. Mater. 10: e2101329.

Nuhn, H. and H.-A. Klok. 2008. Secondary structure formation and LCST behavior of short elastin-like peptides. Biomacromolecules. 9: 2755–2763.

Okada, Y., S. Hosoya, H. Suzuki and K. Chiba. 2014. Total synthesis of elastin peptide using high pressure–liquid phase synthesis assisted by a soluble tag strategy. Org. Lett. 16: 6448–6451.

Partridge, S.M., D.F. Elsden, J. Thomas, A. Dorfman, A. Telser and P.L. Ho. 1966. Incorporation of labelled lysine into the desmosine cross-bridges in elastin. Nature. 209: 399–400.

Patel, J., H. Zhu, R. Menassa, L. Gyenis, A. Richman and J. Brandle. 2007. Elastin-like polypeptide fusions enhance the accumulation of recombinant proteins in tobacco leaves. Transgenic Res. 16: 239–249.

Petitdemange, R., E. Garanger, L. Bataille, K. Bathany, B. Garbay, T.J. Deming, et al. 2017a. Tuning thermoresponsive properties of cationic elastin-like polypeptides by varying counterions and side-chains. Bioconjug. Chem. 28: 1403–1412.

Petitdemange, R., E. Garanger, L. Bataille, W. Dieryck, K. Bathany, B. Garbay, et al. 2017b. Selective tuning of elastin-like polypeptide properties via methionine oxidation. Biomacromolecules. 18: 544–550.

Pramounmat, N., S. Asaei, J.D. Hostert, K. Young, H.A. von Recum and J.N. Renner. 2022. Grafting of short elastin-like peptides using an electric field. Sci. Rep. 12: 18682.

Prhashanna, A., P.A. Taylor., J. Qin, K.L. Kiick and A. Jayaraman. 2019. Effect of peptide sequence on the LCST-Like transition of elastin-like peptides and elastin-like peptide-collagen-like peptide conjugates: Simulations and experiments. Biomacromolecules. 20: 1178–1189.

Roberts, S., M. Dzuricky and A. Chilkoti. 2015. Elastin-like polypeptides as models of intrinsically disordered proteins. FEBS Letters. 589: 2477–2486.

Rodríguez-Cabello, J.C., F.J. Arias, M.A. Rodrigo and A. Girotti. 2016. Elastin-like polypeptides in drug delivery. Adv. Drug Deliv. Rev. 97: 85–100.

Rodriguez-Cabello, J.C., I. Gonzalez de Torre, A. Ibanez-Fonseca and M. Alonso. 2018. Bioactive scaffolds based on elastin-like materials for wound healing. Adv. Drug Deliv. Rev. 129: 118–133.

Rong, J., J. Han, Y. Zhou, L. Wang, C. Li and Y. Wang. 2018. Process integration of production, purification, and immobilization of β-Glucosidase by constructing glu-linker-ELP-GB system. Ind. Eng. Chem. Res. 57: 15620–15631.

Shah, M., P.Y. Hsueh, G. Sun, H.Y. Chang, S.M. Janib and J.A. MacKay. 2012. Biodegradation of elastin-like polypeptide nanoparticles. Protein Sci. 21: 743–750.

Sharma, A., P. Sharma and S. Roy. 2021. Elastin-inspired supramolecular hydrogels: a multifaceted extracellular matrix protein in biomedical engineering. Soft Matter. 17: 3266–3290.

Shi, J., Y. Gao, Y. Zhang, Y. Pan and B. Xu. 2011. Calcium ions to cross-link supramolecular nanofibers to tune the elasticity of hydrogels over orders of magnitude. Langmuir. 27: 14425–14431.

Shirazi, A.N., R.K. Tiwari, D. Oh, A. Banerjee, A. Yadav and K. Parang. 2013. Efficient delivery of cell impermeable phosphopeptides by a cyclic peptide amphiphile containing tryptophan and arginine. Mol. Pharmaceutics. 10: 2008–2020.

Strohl, W.R. 2015. Fusion proteins for half-life extension of biologics as a strategy to make biobetters. BioDrugs. 29: 215–239.

Sumiyoshi, S., K. Suyama, D. Tatsubo, N. Tanaka, K. Tomohara, S. Taniguchi, et al. 2022a. Metal ion scavenging activity of elastin-like peptide analogues containing a cadmium ion binding sequence. Sci. Rep. 12: 1861.

Sumiyoshi, S., K. Suyama, N. Tanaka, T. Andoh, A. Nagata, K. Tomohara, et al. 2022b. Development of truncated elastin-like peptide analogues with improved temperature-response and self-assembling properties. Sci. Rep. 12: 19414.

Suyama, K., S. Taniguchi, D. Tatsubo, I, Maeda and T. Nose. 2016. Dimerization effects on coacervation property of an elastin-derived synthetic peptide $(FPGVG)_5$. J. Pept. Sci. 22: 236–243.

Suyama, K., D. Tatsubo, W. Iwasaki, M. Miyazaki, Y. Kiyota, I. Takahashi, et al. 2018. Enhancement of self-aggregation properties of linear elastin-derived short peptides by simple cyclization: Strong self-aggregation properties of cyclo[FPGVG]$_n$, consisting only of natural amino acids. Biomacromolecules. 19: 3201–3211.

Suyama, K., M. Mawatari, D. Tatsubo, I. Maeda and T. Nose. 2021. Simple regulation of the self-assembling ability by multimerization of elastin-derived peptide (FPGVG)$_n$ using nitrilotriacetic acid as a building block. ACS Omega. 6: 5705–5716.

Suyama, K., M. Shimizu, I. Maeda and T. Nose. 2022. Flexible customization of the self-assembling abilities of short elastin-like peptide Fn analogs by substituting N-terminal amino acids. Biopolymers. e23521.

Suyama, K., M. Murashima, I. Maeda and T. Nose. 2023. Enhancement of aggregate formation through aromatic compound adsorption in elastin-like peptide (FPGVG)$_5$ analogs. Biomacromolecules. 24: 5265–5376.

Tanaka, N., K. Suyama, K. Tomohara, I. Maeda and T. Nose. 2023. Branched short elastin-like peptides with temperature responsiveness obtained by EDTA-mediated multimerization. J. Pept. Sci. 29: e3449.

Tatsubo, D., K. Suyama, M. Miyazaki, I. Maeda and T. Nose. 2018. Stepwise Mechanism of temperature-dependent coacervation of the elastin-like peptide analogue dimer, (C(WPGVG)$_3$)$_2$. Biochemistry. 57: 1582–1590.

Taylor, P.A., H. Huang, K.L. Kiick and A. Jayaraman. 2020. Placement of tyrosine residues as a design element for tuning the phase transition of elastin-peptide-containing conjugates: Experiments and simulations. Mol. Syst. Des. Eng. 5: 1239–1254.

Urry, D.W., T.L. Trapane, H. Sugano and K.U. Prasad. 1981. Sequential polypeptides of elastin: Cyclic conformational correlates of the linear polypentapeptide. J. Am. Chem. Soc. 103: 2080–2089.

Urry, D.W. 1988. Entropic elastic processes in protein mechanisms. I. Elastic structure due to an inverse temperature transition and elasticity due to internal chain dynamics. J. Protein Chem. 7: 1–34.

Urry, D.W., D.K. Chang, N.R. Krishna, D.H. Huang, T.L. Trapane and K.U. Prasad. 1989. Two-dimensional proton NMR studies on poly(VPGVG) and its cyclic conformational correlate, cyclo-(VPGVG)$_3$. Biopolymers. 28: 819–833.

Urry, D.W., C.H. Luan, T.M. Parker, D.C. Gowda, K.U. Prasad, M.C. Reid, et al. 1991a. Temperature of polypeptide inverse temperature transition depends on mean residue hydrophobicity. J. Am. Chem. Soc. 113. 4346–4348.

Urry, D.W., T.M. Parker, M.C. Reid and D.C. Gowda. 1991b. Biocompatibility of the bioelastic materials, poly(GVGVP) and its γ-irradiation cross-linked matrix: Summary of generic biological test results. J. Bioact. Compat. Polym. 6: 263–282.

Urry, D.W. 1992. Free energy transduction in polypeptides and proteins based on inverse temperature transitions. Prog. Biophys. Mol. Biol. 57: 23–57.

Urry, D.W., D.C. Gowda, T.M. Parker, C.-H. Luan, M.C. Reid, C.M. Harris, et al. 1992. Hydrophobicity scale for proteins based on inverse temperature transitions. Biopolymers. 32: 1243–1250.

Urry, D.W. 1997. Physical chemistry of biological free energy transduction as demonstrated by elastic protein-based polymers. J. Phys. Chem. B 101: 11007–11028.

Urry, D.W. and A. Pattanaik. 1997. Elastic protein-based materials in tissue reconstruction. Ann. N.Y. Acad. Sci. 831: 32–46.

Urry, D.W., K.D. Urry, W. Szaflarski and M. Nowicki. 2010. Elastic contractile model proteins: Physical chemistry, protein function and drug design and delivery. Adv. Drug Delivery Rev. 62: 1404–1455.

Varanko, A.K., J.C. Su and A. Chilkoti. 2020. Elastin-like polypeptides for biomedical applications. Annu. Rev. Biomed. Eng. 4: 343–369.

Vasconcelos, A., A.C. Gomes and A. Cavaco-Paulo. 2012. Novel silk fibroin/elastin wound dressings. Acta Biomater. 8: 3049–3060.

Vrhovski, B. and A.S. Weiss. 1998. Biochemistry of tropoelastin. Eur. J. Biochem. 258: 1–18.

Wang, Z., Y. Li, Y. Huang, M.P. Thompson, C.L. LeGuyader, S. Sahu, et al. 2015. Enzyme-regulated topology of a cyclic peptide brush polymer for tuning assembly. Chem. Commun. 51: 17108–17111.

Wang, B., R. Pan, W. Zhu, Y. Xu, Y. Tian, M. Endo, et al. 2021. Short intrinsically disordered polypeptide-oligonucleotide conjugates for programmed self-assembly of nanospheres with temperature-dependent size controllability. Soft Matter 17: 1184–1188.

Wood, D.W., W. Wu, G. Belfort, V. Derbyshire and M. Belfort. 1999. A genetic system yields self-cleaving inteins for bioseparations. Nat. Biotechnol. 17: 889–892.

Wright, E.R., R.A. McMillan, A. Cooper, R.P. Apkarian and V.P. Conticello. 2002. Thermoplastic elastomer hydrogels via self-assembly of an elastin-mimetic triblock polypeptide. Adv. Funct. Mater. 12: 149–154.

Wu, X., R. Sallach, C.A. Haller, J.A. Caves, K. Nagapudi, V.P. Conticello, et al. 2005. Alterations in physical cross-linking modulate mechanical properties of two-phase protein polymer networks. Biomacromolecules. 6: 3037–3044.

Yang, J., Z. Wang, X. Liang, W. Wang and S. Wang. 2024. Multifunctional polypeptide-based hydrogel bio-adhesives with pro-healing activities and their working principles. Adv. Colloid Interface Sci. 327: 103155.

Yeo, G.C., F.W. Keeley and A.S. Weiss. 2011. Coacervation of tropoelastin. Adv. Colloid Interface Sci. 167: 94–103.

Zhang, Y.-N., R.K. Avery, Q. Vallmajo-Martin, A. Assmann, A. Vegh, A. Memic, et al. 2015. A highly elastic and rapidly crosslinkable elastin-like polypeptide-based hydrogel for biomedical applications. Adv. Funct. Mater. 25: 4814–4826.

Zhao, H., E. Ibarboure, V. Ibrahimova, Y. Xiao, E. Garanger and S. Lecommandoux. 2021. Spatiotemporal dynamic assembly/disassembly of organelle-mimics based on intrinsically disordered protein-polymer conjugates. Adv. Sci. 8: 2102508.

Zhou M., Y. Shmidov, J.B. Matson and R. Bitton. 2017. Multi-scale characterization of thermo-responsive dendritic elastin-like peptides. Colloids Surf. B. Biointerfaces. 153: 141–151.

Chapter 8

Unconventional Fluorescence Properties of Non-aromatic Amino Acids: Material and Biological Applications

Vishnu N. Vijayan[1], Karthika Kannan[1]
and Sushabhan Sadhukhan*[1,2]

[1]Department of Chemistry, Physical and Chemical Biology Laboratory, Indian Institute of Technology Palakkad, Kerala 678 623, India.

[2]Department of Biological Sciences and Engineering, Indian Institute of Technology Palakkad, Kerala 678 623, India.

INTRODUCTION

Amino acids exhibit fluorescence emission when they absorb light typically in the ultraviolet range. This property is particularly significant in aromatic amino acids like tryptophan, tyrosine, and phenylalanine (Figure 8.1a), which are commonly used as intrinsic fluorescent probes in biochemical studies to monitor protein folding, structure, dynamics, and interaction (Liao et al. 2021, Tomalia et al. 2019). The presence of aromatic groups in a compound significantly contribute to the

*For Correspondence: sushabhan@iitpkd.ac.in

fluorescence emission due to the low energy π-π^* electronic transition. Emission from aromatic biomolecules (Gu et al. 2018) and aromatic amino acids such as tyrosine and tryptophan have been a well-known phenomenon reported by several research groups (Rehms and Callis 1993, Trubetskaya et al. 2016). The strong fluorescence from aromatic and organic luminophores led to various applications of these molecules in the field of Organic Light Emitting Diodes (OLEDS), chemo sensors and biosensors, bioimaging etc., (Gao et al. 2023). Traditionally, fluorescence has been primarily associated with conjugated systems, which exhibit strong emission due to their delocalized π-electron systems. However, recent advances have shown that non-aromatic molecules such as non-aromatic amino acids, under certain conditions, can also exhibit fluorescence, often this fluorescence emission can be explained by Clustering-Triggered Emission (CTE) mechanism. When subjected to 365 nm UV irradiation, a variety of non-aromatic amino acids with different side chains demonstrated significant increases in their emission intensity as evident from the quantum yield values (Chen et al. 2018a) (Figure 8.1b). This enhanced emission performance was clearly established by analyzing their recrystallized solids, which revealed a substantial link between molecular arrangement and improved emission characteristics. These findings indicate that the structural configuration of amino acids is critical in improving their photophysical performance under UV illumination. Despite the potential applicability of fluorescence from traditional conjugated systems in various fields, several shortcomings of the organic luminophores in terms of their synthesis and structure hinder their extended applications. Aside from their widespread use in biology, aromatic amino acids like tryptophan, tyrosine, and phenylalanine have some limitations because, their fluorescence is highly sensitive to environmental factors like pH, polarity, and other quenching factors like oxygen or metal ions. Tyrosine and phenylalanine have even lower environmental sensitivity, making them unsuitable for investigating detailed structural changes in proteins or complexes. Furthermore, aromatic amino acids are susceptible to photobleaching, which limits their long-term usage in fluorescence microscopy research (Ghisaidoobe and Chung 2014). Thus the shift from aromatic to non-aromatic fluorescence expands the scope of fluorescence applications, allowing for the design of new fluorescent materials and probes that do not rely on aromatic structures.

Previous reports suggested that protein fluorescence basically originated due to the aromatic side chains (Weinryb and Steiner 1970, Brun et al. 1975, Chen and Barkley 1998, Samanta et al. 2000, Xu et al. 2020). However, the first report in 1970 on blue luminescence from a saturated amine named 1,4-diazabicyclo[2.2.2]octane (DABCO) resulted in a paradigm shift towards the non-traditional fluorescence phenomenon, which does not meet the above-mentioned criteria for exhibiting fluorescence (Halpern 1970). In this context, by defying the concept of aromatic fluorescence, the emission from non-aromatic counterparts of the biomolecules has been reported. Probably less attention has been paid to the non-aromatic counterparts as they lack conventional chromophores in them. Homchaudhuri and Swaminath (2001) serendipitously discovered intrinsic luminescence from lysine, a non-aromatic amino acid in 2001. The term 'nonconventional chromophores' is used to distinguish luminophores from conventional building blocks with aromatic groups or bulky conjugates. Non-conventional chromophores, unlike traditional building blocks, lack cumbersome conjugated structures. Over the years, many research groups tried to explain the mechanism behind the non-conventional fluorescence with several conjectures such as photoinduced charge transfer (Alom and Swaminathan 2023), oxidation (Wang and Imae 2004, Wang et al. 2007), carbonyl gathering (Zuo et al. 2021) etc. Although some specific properties of this unusual fluorescence were disclosed by these approaches, other discoveries remain unanswered. Chen et al. reported the emission behavior of non-conventional luminophores present in a system as CTE (Chen et al. 2018a). In another report by Wang et al., an intrinsic fluorescence emission from common rice, and a deep investigation of starch structures revealed the presence of 'clustered chromophores' containing electron-rich oxygen atoms that were presumed to be responsible for this emission. Molecules possessing non-conjugation, saturated groups, and heteroatoms (e.g., N, O, P, S) emits in concentrated solutions, which forms the basis of CTE. Other characteristics of CTE include concentration-enhanced luminescence, excitation wavelength dependence, and room temperature phosphorescence. The reason for room temperature phosphorescence is also interesting, as the CTE molecules have heteroatoms, and cause spin-orbit coupling, thereby promoting intersystem crossing (Liao et al. 2021).

The fluorescence originating from the non-aromatic amino acids arises due to groups such as NH_2, C=O, COOH, OH, etc (Figure 8.1c). Clustering of these unconventional chromophores in solution, when they are in close proximity, leads to the structural rearrangement of the atoms or the groups present. The subsequent Through-Space Conjugation (TSC) followed by the conformation rigidification leads to enhanced fluorescence emission behavior. Particularly the C=O oxygen atoms is suspected to have a prominent role in the interactions due to the electron delocalization ability and interaction with the nearby atoms (Niyangoda et al. 2017). Thus, to explain the intrinsic fluorescence seen in proteins, the CTE mechanism can be applied and in the interim, is useful for learning about the applicability in material and biological systems. A large number of systems showing this non-traditional fluorescence behavior are being developed from the concept of CTE

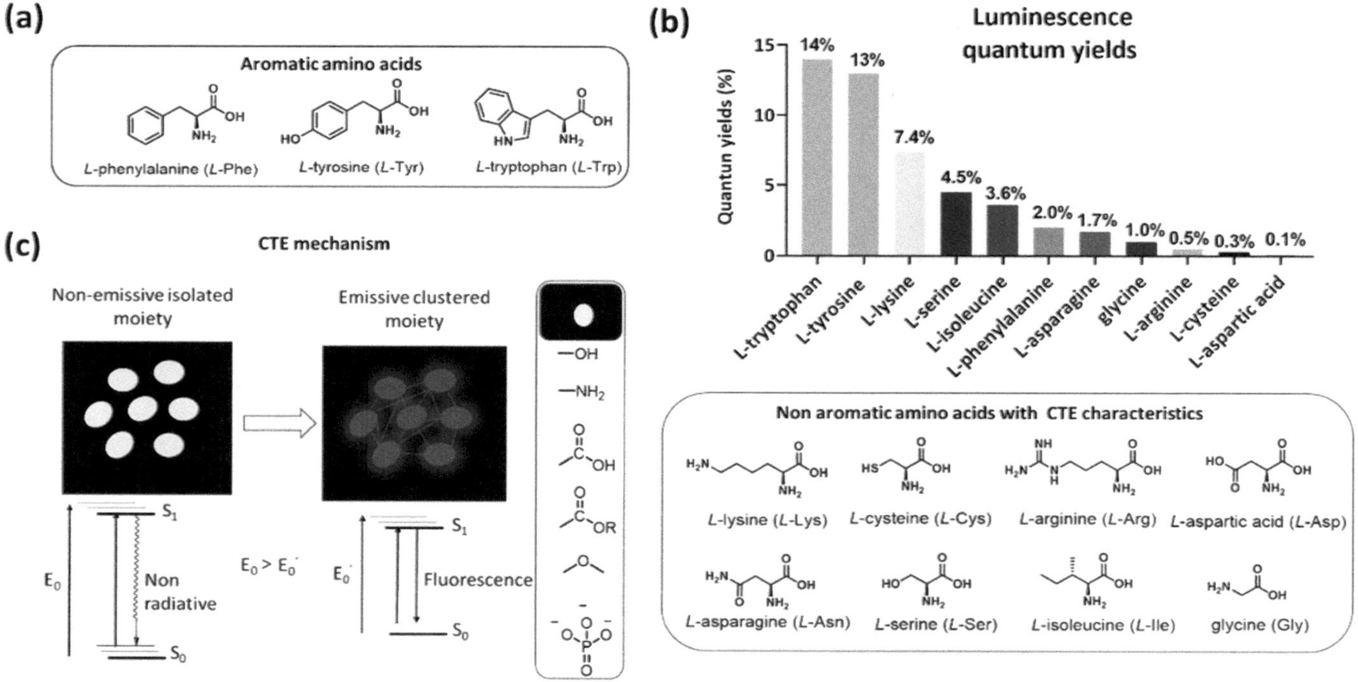

Figure 8.1 (a) Chemical structures of aromatic amino acids. (b) Bar diagram showing the quantum yields of aromatic and non-aromatic amino acids under 365 nm UV excitation and their chemical structures. (c) Schematic illustration of CTE mechanism.

mechanism such as biomacromolecules (Wang et al. 2019), polymeric systems that are both linear and hyperbranched (Chen et al. 2018b, Wang et al. 2007) supramolecular structures, (Liu et al. 2016) etc. These structures generally contain heteroatoms and electron rich moieties that augment the emission via through-space conjugation and subsequent conformation rigidification. There are reports that support the application of non-aromatic biomolecules in mammalian and bacterial cell imaging (Bonifácio et al. 2012, Chen et al. 2018c, Wang et al. 2015, Xu et al. 2020). Hence, here we have tried to integrate the knowledge of the non-aromatic fluorescence from amino acids and related systems to provide an overview of this intriguing emission phenomenon.

MECHANISM OF CTE

The mechanism of non-traditional luminescence is highly debated, with various theories proposed. Different systems have led to differing explanations, and even within the same system, consensus is difficult to achieve. However, one of the most promising mechanisms which probably leads to CTE was proposed by Yang et al. in 2013 where they have observed luminescence emission from common rice, on investigation of the main components, such as starch and other natural product, blue luminescence emission was observed even in the absence of conventional fluorophores in it. Given their structures, "clustered chromophores" composed of electron-rich groups, such as oxygen atoms, are believed to be responsible for light emission. In an isolated system or in dilute solutions, the molecules are far apart, which makes it difficult for the molecule to get excited to higher excited states by appropriate energy activation and the non-radiative relaxation rates are fast due to high rotational and vibrational motions, therefore no visible emission can be observed for these molecules. However, for the clustered chromophores the molecules are in close proximity to each other, where the atomic distances are smaller than the van der Waals radius, which results in effective interactions. This causes the energy gap to be narrowed and extended electron delocalisation is exhibited by these clustered chromophores. Conformational rigidification, enabled by strong intramolecular and intermolecular interactions, significantly hinders non-radiative deactivations, allowing the clustered chromophores to emit light on excitation. Later in 2016, this mechanism was termed as CTE (Zhou et al. 2016). In the context of CTE, the Jablonski diagram can be used to demonstrate clustering behavior. The fundamental distinctions are how clustering affects electronic interactions and non-radiative decay mechanisms (Figure 8.2a). Non-bonding interactions are particularly probable in systems with electron-rich atoms when

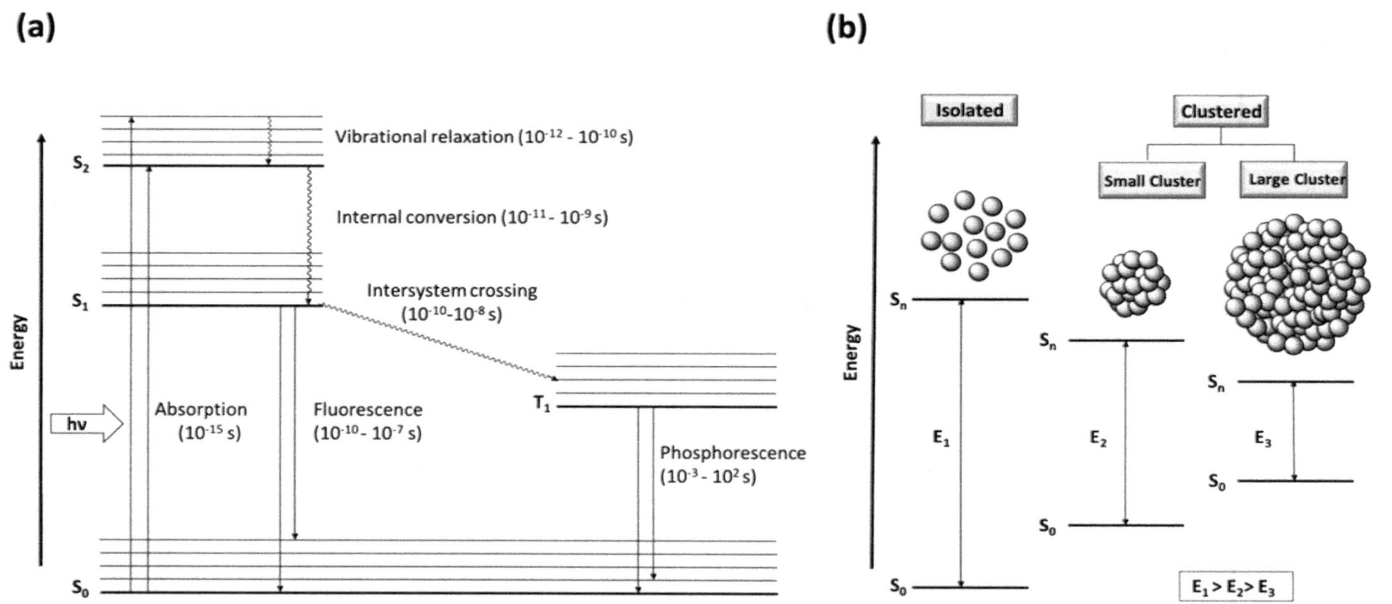

Figure 8.2 (a) Typical Jablonski diagram showing electronic transitions and the time-scale of the respective photophysical processes are shown in the bracket. (b) Cluster size-based CTE emission mechanism.

the distance between the two atoms is within a favorable range. In such instances, the electron density around these atoms can lead to a variety of non-covalent interactions, including van der Waals forces, dipole-dipole interactions, and even hydrogen bonding, providing the atoms are properly positioned. Zhang et al. (2020) developed the Through Space Conjugation (TSC) model for explaining the strong non-bonding interactions occurring due to the clustering of atoms when they are in proximity. The size of the clustered atoms can significantly influence the emission properties arising in the CTE system (Figure 8.2b). As the molecules get converted from isolated to clustered forms emission can vary within the small and large clusters formed. The energy gap can significantly reduce as the cluster size increases, intra or intermolecular TSC forms the basis of this clusteroluminescence (Zhang et al. 2020). Single crystal analysis of D-(+)-glucose supports the observation revealing effective intra and intermolecular interactions within its crystals. These interactions create a 3D network of O···O short contacts, forming oxygen clusters with enhanced electron delocalization and rigidified conformations (Zhou et al. 2020). Thus, CTE can be considered as an emission arising from the non-conventional chromophores due to the presence of *n* and *π* electrons present in a system and the corresponding conformation rigidification, which results in the effective through-space conjugation leading to enhanced emission.

The non-conventional chromophores can be heteroatoms atoms, such as Nitrogen (N) (Wang and Imae 2004), Sulphur (S) (Yan et al. 2015), Oxygen (O) (Gong et al. 2013), and Phosphorous (P) (Bhattacharya et al. 2012), which contain free lone pairs of electrons that can enhance the through-space electronic interactions, when they are in close proximity in concentrated solutions. These elements originate from functional groups such as hydroxyl (–OH), amine ($–NH_2$), carboxyl (–COOH), ester (–COOR), ether (–O–), phosphate ($–PO_4^{3-}$), etc. Another important characteristic of CTE materials as a result of TSC is the excitation-dependent emission phenomena. In a CTE system, the clusters formed can vary greatly in size and structure, leading to diverse spatial environments for each clusteroluminogen. Consequently, the interactions between these clusteroluminogen differ, making each one a distinct emissive species within the system, characterized by unique excitation and emission wavelengths. In addition to the hydrophobic effect and structural entanglement, intermolecular and intramolecular electrostatic interactions from dipoles or transient dipoles contribute to the development of clusteroluminogens (Wang et al. 2015). Chen et al. provided a comprehensive study on the emission mechanism from non-aromatic amino acids and poly (amino acids). The experimental results showed that non-aromatic amino acids and poly (amino acids) have considerable intrinsic fluorescence, as seen by emission in the UV and visible regions at certain excitation wavelengths. Spectroscopic examination, including steady-state fluorescence and time-resolved observations, indicated that the emission comes from backbone structures and side-chain interactions, not from aromatic residues. These findings demonstrate that the intrinsic emission characteristics of non-aromatic amino acids may be systematically controlled by parameters like pH, solvent polarity, and peptide conformation, laying the groundwork for their usage in fluorescence-based applications like bioimaging and molecular probes (Chen et al. 2018c). The excitation-dependent emission characteristic of non-aromatic amino acid, L-lysine was reported by Sahadevan et al. where a shift of emission peak of L-lysine to longer wavelengths can be visualized by varying the excitation wavelengths (Sahadevan et al. 2023).

Molecular level description of non-aromatic fluorescence has been explained by various theoretical calculations and simulations. To study the non-equilibrium photoemission process, researchers used Molecular Dynamics (MD) simulations, Time-Dependent Density Functional Theory (TDDFT), and Algebraic Diagrammatic Construction (ADC)-based electronic structure calculations. Furthermore, *ab initio* nonadiabatic molecular dynamics simulations are used to investigate these phenomena (Pinotsi et al. 2016). When we look at protein structure, proteins absorb and emit UV light by a $\pi \rightarrow \pi^*$ transition on the side chains of three aromatic amino acids, resulting in emission at 350 nm (3.5 eV). The amide bonds produce mild $n \rightarrow \pi^*$ transitions at 225 nm (5.5 eV) and powerful $\pi \rightarrow \pi^*$ excitations about 207 nm (6 eV) (Grisanti et al. 2020) The origin of the unexpected red shift in non-aromatic emission spectra, relative to conventional $\pi \rightarrow \pi^*$, $n \rightarrow \pi^*$, or charge transfer transitions, remains unknown. The distinctive photophysics of non-aromatic fluorescent systems may be attributable to the close proximity of peptide bonds in amyloid-like aggregates, which is assisted via hydrogen bond exchange (Shukla et al. 2004, Del Mercato et al. 2007). One of the initial hypotheses proposed that hydrogen bonds between the –CO and –NH groups of peptide bonds could contribute to partial electron delocalization and induce structural rearrangements. These changes might lead to lower energy electronic transitions and enhanced radiative recombination efficiency. In particular, hydrogen bonds can stabilize excited states, where negative charge accumulates on the hydrogen bond acceptor groups, resulting in a red shift in excitation energies. Supporting this idea, studies by Pinotsi et al. demonstrated that fully protonated or deprotonated amyloid fibrils exhibit a significant decrease in fluorescence intensity. This observation suggests that proton delocalization along the hydrogen bonding network plays a crucial role in enhancing the fluorescence quantum yield. The delocalized protons within the (Hydrogen Bonds) HBs potentially facilitate more efficient recombination of excited electrons with holes, thereby boosting the overall fluorescence emission. These findings point to a mechanism in which hydrogen bonding interactions and proton dynamics within amyloid structures modulate both the energy of electronic transitions and the efficiency of light emission in non-aromatic fluorescent systems (Pinotsi et al. 2016). Later, Schiattarella et al. discovered that amyloid fibrils with protonated or zwitterionic ends had equal fluorescence intensity, questioning some of the previous conclusions about the role of protonation in fluorescence modulation (Schiattarella et al. 2022).

Grisanti et al. (2020) recently proposed an alternative molecular interpretation for the near-UV and visible emission energies observed in non-aromatic systems. Their study suggests that when the molecular geometries of amides within amyloid proteins deviate from planarity, there is a significant elongation of the CO and CN bonds in the amide groups. This structural distortion causes a slight destabilization of the ground state while considerably stabilizing the $n\pi^*$ excited states by more than 1.0 eV, effectively shifting the emission toward the visible range. In addition to the red shift, the $n\pi^*$ states, which are typically considered "dark" due to their low emission probability, exhibit a significant increase in emission efficiency. This enhancement, by an order of magnitude, is attributed to the coupling of the $n\pi^*$ states with vibrational modes associated with the deplanarization of the amide bonds. This coupling increases the likelihood of radiative transitions, allowing these previously low-emission states to contribute more effectively to the fluorescence observed in non-aromatic systems. This alternative mechanism provides valuable insight into how molecular geometry and vibrational coupling can influence the photophysical properties of amyloid aggregates (Grisanti et al. 2020). Based on a number of investigations conducted by a research group, a plausible theory has emerged that shows certain chemical interactions-like proton transfer along short hydrogen bonds or vibrational modes involving the carbonyl stretch-have a major impact on fluorescence behavior (Stephens et al. 2021, Pinotsi et al. 2016). The vibrational potentials for carbonyl stretches can be weakened by environmental factors and proton transfer along short hydrogen bonds, which can increase the energy barriers for achieving conical junctions. Although the relationship between non-aromatic fluorescence and HBs has been established, the specific type of chromophore that is responsible for this emission is yet unknown. Carbonyl-containing organic solvents were found to display intrinsic blue fluorescence in a study by Niyangoda et al., suggesting that carbonyl groups could be the main source of emission in both scenarios (Niyangoda et al. 2017). Grisanti et al.'s computational investigations have demonstrated that the most likely emission states in non-aromatic amyloids are n–π^* electronic transitions focused on carbonyl oxygen atoms, which lends support to this theory (Grisanti et al. 2020). These results imply that the carbonyl groups are essential to the fluorescence seen in these systems, especially when it comes to non-aromatic structures like amyloids, as they participate in n-π^* transitions. This theoretical understanding offers a plausible explanation for the emission mechanism; nonetheless, additional experimental confirmation is necessary to completely clarify the identity of the chromophore.

Another possible explanation for the UV-visible absorption and fluorescence seen in proteins devoid of aromatic amino acids is charge transfer transitions followed by electron-hole recombination (Prasad et al. 2017, Kumar et al. 2020). This implies that different electronic processes might be at work to achieve these optical properties in the absence of aromatic side chains, which are generally recognized for their potent UV absorption and fluorescence. In particular, excited states may result from charge transfer between various functional groups within the protein structure; these states then relax through electron-hole recombination to produce the observable fluorescence. Strong absorption has been found in the 250–800 nm wavelength region for both concentrated solutions enriched in charged amino acids and synthetic monomeric proteins. Protein charge transfer spectra (ProCharTS) is the term used to describe this absorption phenomena where the underlying mechanism of these charge transfer changes has been revealed by recent atomistic simulations that combine static time-dependent density functional theory calculations with classical molecular dynamics. The simulations demonstrate that in addition to contributions from the peptide backbone, the charge transfer is mediated by interactions between the amino (–NH_3^+) and carboxylate (–COO^-) groups present in the side chains of charged amino acids. The charged moieties have a significant role in the electronic transitions that give rise to the absorption spectra that are observed. According to the study, charge transfer between these polar functional groups inside the protein structure happens as a result of the charged fragments' closeness and orientation. As a result, the network of extended charge delocalization is created, which adds to ProCharTS's wide absorption range (Prasad et al. 2017). The interaction between the backbone and the charged side chains sheds more light on how protein architecture affects charge transfer dynamics and offers a better understanding of the electronic characteristics of biomolecules. Room temperature phosphorescence emission from proteins were reported initially by Debye and Edwards (1952). This foundational study initiated the research in protein phosphorescence, highlighting that intrinsic phosphorescence mainly arises from aromatic residues like tyrosine, which serves as an intrinsic phosphorescent probe. Early research focused primarily on dilute solutions, while the solid-state phosphorescence of biomolecules was rarely explored. However, the

CTE mechanism, revealed that many biomolecules, even without traditional chromophores, exhibited strong emissions in aggregated states. The CTE mechanism further suggests that photoluminescence is no longer exclusive to aromatic compounds, opening new possibilities for discovering or designing unconventional luminophores.

NON-AROMATIC INTRINSIC EMISSION FROM BIOMOLECULES

Non-aromatic Amino Acids

Proteinogenic amino acids are essential to all living organisms. Among them, aromatic amino acids, including phenylalanine, tryptophan, and tyrosine, which are intrinsically luminescent, are thought to contribute to the luminescence observed in proteins (Wang et al. 2015). In this context, little attention was paid to emission characteristics of non-aromatic amino acids as they lack a well-defined chromophoric unit. However, several research groups discussed the emission from non-aromatic amino acids when they are in solution form (Homchaudhuri and Swaminathan 2001). Dilute aqueous solutions of non-aromatic amino acids are non-emissive, but significant visible emissions are observed in most of their concentrated forms. Homchaudhuri and Swaminathan (2004) studied the absorption and fluorescence properties of concentrated aqueous solution of L-lysine, observing blue fluorescence (~435 nm) when excited at 355 nm. They also found that L-lysine monohydrochloride showed enhanced fluorescence at higher concentrations, likely due to the aggregation of side amino groups. The molecular structure of L-lysine includes electron-rich α-carboxyl, α-amino, and side amino groups. When these groups are closely gathered, extended electronic delocalization likely forms clustered chromophores responsible for the observed blue luminescence. Further investigation on ε-poly-L-lysine which is a biodegradable polypeptide that contains 25-35 L-lysine residues revealed that they also showed highly luminescent behavior (Homchaudhuri and Swaminathan 2004). Arnon et al. (2021) conducted a systematic study on natural amino acids, discovering that visible fluorescence in amino acid crystals is packing-dependent, including both aromatic and non-aromatic types. Charged non-aromatic amino acids, especially lysine, showed the strongest fluorescence signals. Notably, cysteine exhibited crystal structure-dependent fluorescence, with a reversible on-off optical transition between two different crystal arrangements (Arnon et al. 2021). This study highlights the significant impact of supramolecular assembly on the optical properties of amino acids. Glycine which is the simplest amino acid itself lacks fluorescence, Cossu et al. (2022) demonstrated that on thermal polymerization, chromophores are generated which results in the formation of fluorescent glycine structures. The use of boric acid resulted in the amidation reaction and subsequently resulted in the formation of polymerized glycine with emissive fluorophores (Cossu et al. 2022). Non-aromatic amino acids such as L-serine, L-lysine, L-asparagine, L-aspartic acid, etc., in solid crystalline forms were found to be emissive due to the clustering of electron-rich moieties (Chen et al. 2018a). In a work done by Li et al. (2023) unique clusteroluminogens have been prepared by modifying the cyclodextrin (CD) with various non-aromatic amino acids such as glycine, isoleucine, methionine, cysteine, glutamic acid, glutamine, asparagine, lysine, and arginine. These nonconventional luminophores, specifically amino acids, are chemically grafted onto CD instead of forming an inclusion complex. This covalent attachment increases the structural rigidity of the CD, facilitating the intermolecular interactions required for the CTE effect. The unusual fluorescence of CD amino acids can be due to increased intermolecular and intramolecular clustering of chromophores, such as amine (–NH2) groups, carboxyl (–COOH), hydroxyl (–OH), and, within the limited area of CD (Li et al. 2023). This clustering effect improves the interaction of chromophores, resulting in distinct photophysical properties. Simultaneously, the rigidification of molecular conformations caused by the creation of many hydrogen bonds strengthens and stabilizes these chromophores. This conformational limitation inhibits non-radiative decay pathways, which contributes to the increased fluorescence. The combination of clusterization and conformational rigidification is critical in the fluorescence mechanism of CD amino acids, resulting in their unique optical characteristics.

Non-aromatic amino acids, including L-glutamine, L-isoleucine, L-asparagine, L-valine, L-threonine, and L-methionine, show bright fluorescence when crystallized in a solid state. Crystallization of natural amino acids was done by antisolvent crystallization method and careful studies of their unit cells and macrostructures with modern techniques such as single-crystal X-ray crystallography, X-ray powder diffraction (XRD), and Scanning Electron Microscopy (SEM) were carried out. These methods provided us with a thorough understanding of the structural characteristics of amino acid crystals. The fluorescence lifetimes of amino acids varied due to changes in hydrogen-bonding network density and the presence of π–π and n–π interactions. The fluorescence behavior appears to be controlled by the amino acid crystal's unique macrostructures, showing the complex link between the crystal structure and photophysical properties. This work reveals that elements such as molecular packing and intermolecular forces within the crystals play important roles (Ravanfar et al. 2020).

Recent studies have systematically investigated the emissive properties of non-aromatic amino acids, providing deeper insight into their photophysical behavior. These studies revealed that the photophysical properties of these amino acids depend significantly on their physical states. Non-aromatic amino acids were found to be active upon aggregation, showing no emission in dilute aqueous solutions but exhibiting noticeable visible emissions in concentrated solutions and recrystallized solids. At cryogenic temperatures (77 K) concentrated solutions displayed enhanced photoluminescence, while dilute solutions remained non-emissive, suggesting that rigidification alone is insufficient to produce significant luminescence. Crystals of non-aromatic amino acids can exhibit room temperature phosphorescence emissions peaking at 490–500 nm, significantly enhanced and prolonged under cryogenic conditions due to increased conformation rigidification. ε-poly-L-lysine displays notable room temperature phosphorescence due to the effect of the polymer chain and strong intra- and intermolecular interactions (Chen et al. 2018a). Inspired by this intrinsic luminescence, Yuan and colleagues revisited protein luminescence, using Bovine Serum Albumin (BSA) as a model and found that BSA powder also exhibited room temperature phosphorescence at ambient conditions, with prolonged emission in vacuum or when compressed into tablet form (Wang et al. 2019).

Proteins and Polypeptides

Luminescent proteins are widely used in various fields of science enabling visualization of cellular processes as well as for the development of novel biosensors. Compared to modified protein fluorescence, intrinsic luminescence from proteins or polypeptides offers advantages such as ease of use and examining the protein structure (Shukla et al. 2004, Bian et al. 2018), while previous research highlighted the role of aromatic residues in protein luminescence, studies often focused on dilute solutions where aromatic components were crucial. However, in aggregated states, non-aromatic amino acids can also contribute to intrinsic luminescence, as evidenced by reports of intrinsically fluorescent proteins lacking aromatic residues (He et al. 2019). Extensive efforts have been made to reveal the photophysical properties of proteins and peptides in aggregated states. Preliminary mechanisms, primarily involving electron delocalization among peptide bonds due to hydrogen bonding, have been proposed. Shukla et al. first observed violet-blue fluorescence in a series of protein aggregates, including gamma-II crystallin, which emitted visible violet-blue light under UV irradiation (350–390 nm). This discovery led researchers to investigate the emission mechanisms of similar proteins and polypeptides (Shukla et al. 2004). After ruling out other factors, it was determined that the intrinsic fluorescence originated from the aggregated proteins and was independent of aromatic sidechains. Sharpe et al. (2011) further investigated the novel visible fluorescence of an amyloid-like fibril octapeptide (GVGVAGVG) without conventional chromophores. Emission was believed to arise due to the presence of aggregation of β-sheet conformations in the peptide supported by transport of charge via hydrogen bonds (Sharpe et al. 2011). Moreover, to enhance understanding, Grisanti et al. (2017) conducted a computational study on hydrogen bonding interactions in amyloid fibril structures, highlighting the role of hydrogen bonds and proton transfer in visible emission (Grisanti et al. 2017). Additionally, the Gazit laboratory reported blue photoluminescence in peptide nanotubes, attributing it primarily to the unique quantum-confined structure formed by self-assembly (Reches and Gazit 2006).

In 2017, Bhattacharya and colleagues investigated the structural basis of concentration-dependent intrinsic luminescence in Human Serum Albumin (HSA), concluding that the peptide structure, rather than amyloid-like fibrils or specific residues, is responsible for the luminescence (Mukherjee et al. 2017). The photophysical properties of HSA closely resemble those of poly (VGGVG). For instance, HSA fluorescence is both concentration- and excitation-dependent. A dilute HSA solution (~10 μM) shows minimal emission, but as concentration increases to 10–50 μM, PL intensity significantly rises, and the emission peak shifts with changing the excitation wavelength values. Additionally, fluorescence lifetime measurements reveal non-single exponential decay kinetics, attributed to the formation of various β-sheet oligomers, indicating the presence of multiple emissive species in the solution (Sharpe et al. 2011). Recently, Chen et al. (2018b) created a Bovine Serum Albumin (BSA)-based hydrogel with intrinsic fluorescence by denaturing Bovine Serum Albumin (BSA) in a basic environment. Compared to the weak fluorescence of a 20% BSA solution, the hydrogel exhibited significantly enhanced fluorescence. Notably, BSA's intrinsic photoluminescence has been observed in solid states, and a recent study provided deeper insights into its PL behaviors. Dilute aqueous BSA solutions (~0.1 mg/mL) showed almost no emission (Chen et al. 2018b). The bright PL in concentrated BSA solutions and solids is likely due to the clustering of peptide backbones, both aromatic and non-aromatic pendants with π and n electrons, and conformational rigidification. These findings suggest that clustering is a key factor in the PL of nonconventional biomolecular luminophores. Additionally, denser molecular packing helps prevent quenching and non-radiative transitions, resulting in more effective emissions. Deby and Edward (1952) first detected room temperature phosphorescence emission from proteins. This ground-breaking finding sparked protein phosphorescence research, demonstrating that intrinsic phosphorescence is mostly caused by aromatic residues such as tyrosine, which act as intrinsic phosphorescent probes (Deby and Edwards 1952). However, early research concentrated mostly on dilute solutions, with solid-state phosphorescence of biomolecules were not investigated frequently. However, the CTE mechanism demonstrates that many biomolecules, including those without typical chromophores, emit strongly when aggregated.

APPLICATIONS OF INTRINSIC NON-AROMATIC FLUORESCENCE

Material and Biological Applications

The development of fluorescent sensors from non-aromatic moieties represents a significant advancement in the field of material science, bioimaging, and molecular detection (Figure 8.3). The strategic selection of naturally occurring non-aromatic systems that are capable of exhibiting CTE and their applications in sensing has several advantages including bypassing the complicated chemical synthesis and use of hazardous chemicals. More importantly, naturally occurring CTE systems such as non-aromatic amino acids or carbohydrates are environmentally benign to work with. Recently our group has reported the detection of metal ions, such as copper (Cu^{2+}) and cobalt (Co^{2+}) by exploiting the highly blue fluorescence of L-lysine (Sahadevan et al. 2023). In this study, we used the unique features of non-aromatic amino acids to create a highly efficient fluorescence sensor. This is the first time a non-aromatic amino

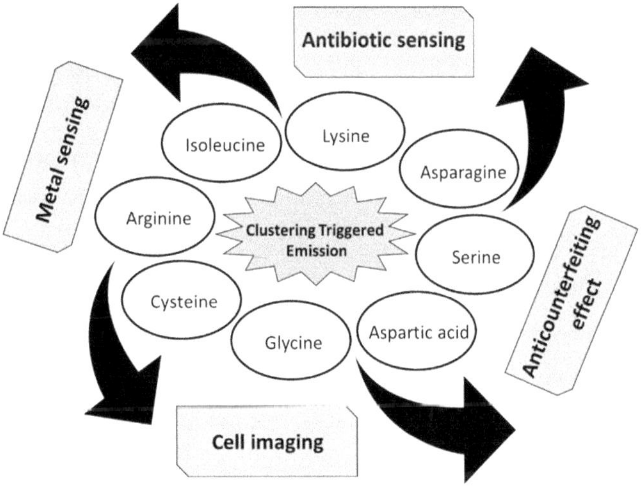

Figure 8.3 Range of material and biological applications of CTE materials.

acid, L-lysine, has been shown to function as a selective and sensitive fluorescence sensor for detecting and distinguishing Cu^{2+} and Co^{2+} ions, colorimetrically. Aqueous solution of L-lysine (6.8 mM) effectively detected Cu^{2+} and Co^{2+} with detection limits of 420 and 460 nM, respectively, in water. Cu^{2+} and Co^{2+} produced different blue and pink colors, allowing for real-time visual investigation. In addition to fluorescence sensing, the system's practical application was shown by developing paper strips that detect Cu^{2+} and Co^{2+}. This sensor performed admirably even in real-world circumstances, reliably detecting these ions in lake and tap water samples. The benefits of this L-lysine based sensor are further underscored by its extraordinarily low cost of the material L-lysine (less than one US cent per 10 mg of L-lysine), reversible functioning where it implies that the addition of EDTA to the solution of L-lysine and Cu^{2+} or Co^{2+} regained the fluorescence of L-lysine solution. Thus, L-lysine could sense Cu^{2+} up to 3 cycles and for Co^{2+} up to 2 cycles without compromising its ability to detect these analytes, and ease of use, making it a remarkably accessible and effective tool for fluorescence-based metal ion detection. As L-lysine was found to be highly fluorescent, later our group thought that ε-poly-L-lysine which is a water-soluble polymer of lysine that contains 25–35 L-lysine residues could have increased photoluminescence properties that can be used for various material applications. As per our anticipation, ε-poly-L-lysine acted as a versatile sensor for the selective detection of only Cu^{2+}. Surprisingly, the limit of detection obtained was found to be as low as 7.11 nM (0.45 ppb), which is much lower than the permissible limit of Cu^{2+} (2 ppm) set by the World Health Organization (WHO) in drinking water samples. CTE features of ε-poly-L-lysine aqueous solution can also be used for anticounterfeiting, preventing unauthorized copying of products. Bright emission of ε-poly-L-lysine can be used to develop anticounterfeiting security signatures or images, a concentrated ε-poly-L-lysine solution (100 mg/mL) was applied on filter paper in different patterns, such as letters or fingerprints, and then dried. Under normal lighting conditions, these markings are imperceptible to the naked eye; nevertheless, when subjected to UV light, the coated areas become noticeable with bright blue fluorescence. UV-induced optical behavior of ε-poly-L-lysine can be used to prevent counterfeiting and forging of the original products thereby increases product security. Moreover, as ε-poly-L-lysine is an FDA-approved food preservative, it can be used for medicinal counterfeiting applications as well. A unique symbol was made on the commercially available paracetamol tablet by using ε-poly-L-lysine that becomes visible only on UV excitation and not in the normal light thus ensuring the authenticity of the tablets being used (Sahadevan et al. 2024). Recently, reports on the applicability of L-lysine to detect specifically the 5-nitroimidazole (5-NI) antibiotics have been developed by Jose et al. (2024). 5-NI drugs, such as ornidazole (ODZ), metronidazole (MDZ), secnidazole (SDZ), and tinidazole (TDZ), are widely used to treat infections such as trichomonal vaginitis, gastroduodenal ulcers, and hypoxic tumors, as well as to promote growth and prevent infection in livestock (Erba and Roche 1978). These medicines are partly absorbed and expelled in urine and feces, and inappropriate disposal can pollute water and soil, creating a serious problem of antibiotic resistance (Sachi et al. 2019, Hassanain et al. 2022). L-lysine can detect these environmentally hazardous antibiotic residues with a limit of detection of 2.5 μM for MDZ, 2.65 μM for ODZ, 3.13 μM for SDZ, and 2.98 μM for TDZ. Practical applicability through paper strips was also feasible by this detection technique making L-lysine a versatile fluorescent sensor (Jose et al. 2024).

Inspired by the bright luminescence behavior of L-lysine, the ε-poly-L-lysine, a biodegradable polypeptide was utilized for the specific detection of 5-NI antibiotics where we could bring down the limit of detection to nanomolar ranging from 3.25–3.97 nM for the evaluated 5-NI antibiotics. Due to its well-known biocompatibility, ε-poly-L-lysine can label cells, as evidenced by the fluorescence emission observed from the cells. The quenching of fluorescence upon the addition of 5-NI antibiotics indicated the application of ε-poly-L-lysine in detecting these antibiotics in the cells (Vijayan et al. 2024). Strong electronegativity of CTE materials allows them to interact with metal ions, altering their aggregation state and light-emitting properties. As a result, they can be used for the detection of metal ions such as Fe^{3+} by their fluorescence quenching (Wang et al. 2018).

Non-aromatic biomolecules are ideal for use in cell or bacteria imaging and as biotags due to their strong intrinsic fluorescence, excellent biocompatibility, low cytotoxicity, and hydrophobic-hydrophilic balance the inherent emission and excellent biocompatibility of non-aromatic amino acids highlight their potential in bioimaging applications. Following a 1.5 h incubation of HeLa cells with 0.1 M L-isoleucine in DMEM, a bright blue fluorescence was detected under confocal microscopy when excited at 405 nm, whereas the control showed no fluorescence. These findings indicate that L-isoleucine is effective for cell staining and, on closer examination, specifically targets endosomes, which is significant for biomedical research and clinical diagnostics (Chen et al. 2018a). Certain biologically relevant structures exhibit non-traditional emission due to the formation of aggregated structures. A notable example is amyloid proteins, where normally soluble proteins aggregate into insoluble fibers that resist cellular repair mechanisms (Rambaran and Serpell 2008). Amyloid fibril formation is related to many diseases such as Parkinson's, Alzheimer's, type 2 diabetes, etc. While all these diseases involve similar types of aggregates, a different protein or peptide is responsible in each case. Intrinsic non-traditional emission provides a non-labeled method for imaging fibril structures. A report by Chan et al. in 2013 revealed that aggregation of several polypeptides including amyloid-β (1–40), amyloid-β (1–42), amyloid-β (33–42), lysozyme and tau into amyloid fibrils resulted in non-traditional emission (Chan et al. 2013). The ability to visualize amyloid structures without external labeling enables new possibilities, particularly for *in vitro* kinetic aggregation assays, where labeling could interfere with the protein aggregation process. On revisiting the fluorescence photo physics of BSA, Wang et al. (2019) suggested that, besides UV emission from Trp residues, BSA produces a strong concentration or aggregation-dependent visible photoluminescence, room temperature phosphorescence with lifetimes up to 34 ms is observed in BSA solids, despite their amorphous form, both visible emission and room temperature phosphorescence were further enhanced under pressurization. The visible emission and room temperature phosphorescence characteristics of BSA solids indicate possible optoelectronic applications. Upon UV light exposure (312 nm), BSA solution on filter paper in the shape of a lotus leaf glows blueish-white. When coated with a polymer, the pattern temporarily glows that lasted for a while following UV treatment in vacuum, but no afterglow is visible when exposed to air. This implies that quenching due to oxygen was responsible for the absence of room temperature phosphorescence in air, leading to the lack of afterglow. This has made the BSA potentially suitable to anti-counterfeiting applications as well as oxygen-sensing agents (Wang et al. 2019).

CONCLUSION AND FUTURE OUTLOOK

Luminophores that lack bulky, aromatic or extensively conjugated chromophores have captured widespread interest because of their fundamental significance and potential applications in optoelectronics and biological fields. Notably, a range of non-aromatic biomolecules, including amino acids, peptides, and polypeptides have been identified as novel emitters. Despite this, the precise emission mechanism underlying their intriguing intrinsic luminescence remains unclear. One of the most important aspects of CTE materials is that they require functional groups with abundant lone pair of electrons. The groups such as hydroxyl (–OH), amine ($-NH_2$), carboxyl (–COOH), ester (–COOR), ether (–O–), phosphate ($-PO_4^{3-}$), etc., which contain electron-rich atoms like O, N, S, P significantly contribute to the CTE characteristics. Excitation-dependent emission and concentration-dependent luminescence enhancement phenomena are profound in CTE materials. Nevertheless, there are certain challenges associated with non-aromatic fluorescence. In many cases, the obtained quantum yields are low compared to the traditional luminophores and this can hamper several practical applications. The emission from these non-aromatic luminophores was also largely restricted to the blue region despite the excitation-dependent emission characteristics. Modifying these non-aromatic molecules with various functional moieties can enhance their emission properties, leading to a red shift in the emission spectrum. Undoubtedly, CTE has a bright future with applications in disciplines as diverse as optoelectronics, bioimaging, and sensing technologies. As researchers continue to investigate and comprehend the mechanisms behind CTE, it is expected that new luminous materials with improved and adjustable characteristics will emerge. These materials may lead to more efficient optoelectronic materials, improved biological probes, and extremely sensitive environmental sensors. The increased demand for eco-friendly and non-toxic alternatives strengthens the case for CTE materials to play an important part in future technology breakthroughs. On a separate note, clustering of amino acid itself can be of immense importance in the context of structure prediction of proteins. In a very recent report, Gupta and Uversky provided a comprehensive understanding on the importance of arginine clusters and its diverse functions in biochemistry (Gupta and Uversky 2024). Clusters of certain amino acids frequently create protein-protein interaction sites, where critical residues (called "hot spots") interact strongly with other proteins. These clusters allow very precise and strong binding interactions that drive biological processes like signal transduction, immunological recognition, and molecular assembly making it a challenging problem to explore in biochemistry, drug design etc.

ACKNOWLEDGMENTS

The authors would like to thank the financial support provided by the Indian Institute of Technology Palakkad. This work was further funded by the Council of Scientific & Industrial Research (CSIR), India (02(0434)/21/EMR-II); the Science and Engineering Research Board (SERB), India (CRG/2023/002841); and Prime Minister's Research Fellowship (PMRF) (ID-3103661).

REFERENCES

Alom, S.E. and R. Swaminathan. 2023. Protein charge transfer spectra in a monomeric protein with no lysine. Phys. Chem. Chem. Phys. 25: 16626–16642.

Arnon, Z.A., T. Kreiser, B. Yakimov, N. Brown, R. Aizen, S. Shaham-Niv, et al. 2021. On-off transition and ultrafast decay of amino acid luminescence driven by modulation of supramolecular packing. iScience. 24: 102695.

Bai, L., H. Yan, T. Bai, Y. Feng, Y. Zhao, Y. Ji, et al. 2019. High fluorescent hyperbranched polysiloxane containing β-cyclodextrin for cell imaging and drug delivery. Biomacromolecules. 20: 4230–4240.

Bhattacharya, S., V.N. Rao, S. Sarkar and R. Shunmugam. 2012. Unusual emission from norbornene derived phosphonate molecule—A sensor for Fe^{III} in aqueous environment. Nanoscale. 4: 6962–6966.

Bian, L., H. Shi, X. Wang, K. Ling, H. Ma, M. Li, et al. 2018. Simultaneously enhancing efficiency and lifetime of utralong organic phosphorescence materials by molecular self-assembly. J. Am. Chem. Soc. 140: 10734–10739.

Bonifácio, V.D.B., V.G. Correia, M.G. Pinho, J.C. Lima and A. Aguiar-Ricardo. 2012. Blue emission of carbamic acid oligooxazoline biotags. Mater. Lett. 81: 205–208.

Brun, F., J.J. Toulmé and C. Hélène. 1975. Interactions of aromatic residues of proteins with nucleic acids. Fluorescence studies of the binding of oligopeptides containing tryptophan and tyrosine residues to polynucleotides. Biochemistry. 14: 558–563.

Chan, F.T.S., G.S. Kaminski Schierle, J.R. Kumita, C.W. Bertoncini, C.M. Dobson and C.F. Kaminski. 2013. Protein amyloids develop an intrinsic fluorescence signature during aggregation. Analyst. 138: 2156–2162.

Chen, Y. and M.D. Barkley. 1998. Toward understanding tryptophan fluorescence in proteins. Biochemistry. 37: 9976–9982.

Chen, X., W. Luo, H. Ma, Q. Peng, W.Z. Yuan and Y. Zhang. 2018a. Prevalent intrinsic emission from nonaromatic amino acids and poly(amino acids). Sci. China Chem. 61: 351–359.

Chen, J., Q. Dong, Y. Huang, X. Ma, T. Fan, Z. Bian, et al. 2018b. Preparation, characterization and application of a protein hydrogel with rapid self-healing and unique auto fluorescent multi-functionalities. J. Biomed. Mater. Res. Part A. 107: 81–91.

Chen, X., X. Liu, J. Lei, L. Xu, Z. Zhao, F. Kausar, et al. 2018c. Synthesis, clustering-triggered emission, explosive detection and cell imaging of nonaromatic polyurethanes. Mol. Syst. Des. Eng. 3: 364–375.

Cossu, F.L., M. Poddighe, L. Stagi, R. Anedda and P. Innocenzi. 2022. The birth of fluorescence from thermally polymerized glycine. Macromol. Chem. Phys. 2200052: 1–8.

Debye, P. and J.O. Edwards. 1952. A note on the phosphorescence of proteins. Science. 116: 143.

Del Mercato, L.L., P.P. Pompa, G. Maruccio, A. Della Torre, S. Sabella, A. Tamburro, et al. 2007. Charge transport and intrinsic fluorescence in amyloid-like fibrils. Proc. Natl. Acad. Sci. USA. 104: 18019–18024.

Du, L.L. B.L. Jiang, X.H. Chen, Y.Z. Wang, L.M. Zou, Y.L. Liu, et al. 2019. Clustering-triggered emission of cellulose and its derivatives. Chinese J. Polym. Sci. 37: 409–415.

Erba, C. and H. Roche. 1978. The nitroimidazole family of drugs. Br. J. Vener. Dis. 54: 69–71.

Ghisaidoobe, A.B.T. and S.J. Chung. 2014. Intrinsic tryptophan fluorescence in the detection and analysis of proteins: A focus on förster resonance energy transfer techniques. Int. J. Mol. Sci. 15: 22518–22538.

Gao, S., S. Wang, J. Wu and Z. Lin. 2023. Regulation and application of organic luminescence from low-dimensional organic-inorganic hybrid metal halides. J. Mater. Chem. C. 11: 16890–16911.

Gong, Y., Y. Tan, J. Mei, Y. Zhang, W. Yuan, Y. Zhang, et al. 2013. Room temperature phosphorescence from natural products: Crystallization matters. Sci. China Chem. 56: 1178–1182.

Grisanti, L., D. Pinotsi, R. Gebauer, G.S. Kaminski Schierle and A.A. Hassanali. 2017. A computational study on how structure influences the optical properties in model crystal structures of amyloid fibrils. Phys. Chem. Chem. Phys. 19: 4030–4040.

Grisanti, L., M. Sapunar, A. Hassanali and N. Došlić. 2020. Toward understanding optical properties of amyloids: A reaction path and nonadiabatic dynamics study. J. Am. Chem. Soc. 142: 18042–18049.

Gu, Y., Z. Zhao, H. Su, P. Zhang, J. Liu, G. Niu, et al. 2018. Exploration of biocompatible AIEgens from natural resources. Chem. Sci. 9: 6497–6502.

Gupta, M.N. and V.N. Uversky. 2024. Reexamining the diverse functions of arginine in biochemistry. Biochem. Biophys. Res. Commun. 705: 149731.

Halpern, A.M. 1970. The vapor state emission from a saturated amine. Chem. Phys. Lett. 6: 296–298.

Hassanain, W.A., C.L. Johnson, K. Faulds, D. Graham and N. Keegan. 2022. Recent advances in antibiotic resistance diagnosis using SERS: focus on the "Big 5" challenges. Analyst. 147: 4674–4700.

He, Z., W. Li, G. Chen, Y. Zhang and W.Z. Yuan. 2019. Polymorphism dependent triplet-involved emissions of a pure organic luminogen. Chinese Chem. Lett. 30: 933–936.

Homchaudhuri, L. and R. Swaminathan. 2001. Novel absorption and fluorescence characteristics of L-lysine. Chem. Lett. 30: 844–845.

Homchaudhuri, L. and R. Swaminathan. 2004. Near ultraviolet absorption arising from lysine residues in close proximity: A probe to monitor protein unfolding and aggregation in lysine-rich proteins. Bull. Chem. Soc. Jpn. 77: 765–769.

Jose, A., V.N. Vijayan, R. Sahadevan, M. Porel and S. Sadhukhan. 2024. Selective luminescent detection of 5-nitroimidazole antibiotics through self-aggregates of a single non-aromatic amino acid, L-lysine. Microchem. J. 197: 109802.

Kumar, A., D. Ahari, A. Priyadarshi, M. Ziauddin Ansari and R. Swaminathan. 2020. Weak intrinsic luminescence in monomeric proteins arising from charge recombination. J. Phys. Chem. B. 124: 2731–2746.

Li, Q., X. Wang, Q. Huang, Z. Li, B.Z. Tang and S. Mao. 2023. Molecular-level enhanced clusterization-triggered emission of nonconventional luminophores in dilute aqueous solution. Nat. Commun. 14: 409.

Liao, P., J. Huang, Y. Yan and B.Z. Tang. 2021. Clusterization-triggered emission (CTE): One for all, all for one. Mater. Chem. Front. 5: 6693–6717.

Liu, C., Q. Cui, J. Wang, Y. Liu and J. Chen. 2016. Autofluorescent micelles self-assembled from an AIE-active luminogen containing an intrinsic unconventional fluorophore. Soft Matter. 12: 4295–4299.

Mukherjee, T.K. 2017. Direct evidence of intrinsic blue fluorescence from oligomeric interfaces of human serum albumin. Langmuir. 33: 10606–10615.

Niyangoda, C., T. Miti, L. Breydo, V. Uversky and M. Muschol. 2017. Carbonyl-based blue autofluorescence of proteins and amino acids. PLoS One, 12: 1–15.

Pinotsi, D., L. Grisanti, P. Mahou, R. Gebauer, C.F. Kaminski, A. Hassanali, G.S. Kaminski Schierle, et al. 2016. Proton transfer and structure-specific fluorescence in hydrogen bond-rich protein structures. J. Am. Chem. Soc. 138: 3046–3057.

Prasad, S., I. Mandal, S. Singh, A. Paul, B. Mandal, R. Venkatramani, et al. 2017. Near UV-visible electronic absorption originating from charged amino acids in a monomeric protein. Chem. Sci. 8: 5416–5433.

Rambaran, R.N. and L.C Serpell. 2008. Amyloid fibrils: Abnormal protein assembly. Prion. 2: 112–117.

Ravanfar, R., C.J Bayles and A. Abbaspourrad. 2020. Structural chemistry enables fluorescence of amino acids in the crystalline solid state. Cryst. Growth De. 20: 1673–1680.

Reches, M. and E. Gazit. 2006. Controlled patterning of aligned self-assembled peptide nanotubes. Nat. Nanotechnol. 1: 195–200.

Rehms, A.A. and P.R. Callis. 1993. Two-photon fluorescence excitation spectra of aromatic amino acids. Chem. Phys. Lett. 208: 276–282.

Sachi, S., J. Ferdous, M.H. Sikder and S.M. Azizul Karim Hussani. 2019. Antibiotic residues in milk: Past, present and future. J. Adv. Vet. Anim. Res. 6: 315–332.

Sahadevan, R., A Jose, M. Porel. and S. Sadhukhan. 2023. Non-aromatic single amino acid, lysine: A simple fluorescence sensor for selective detection of Cu^{2+} and Co^{2+} with colorimetric distinction by the virtue of its self-aggregation behavior. J. Photochem. Photobiol. A Chem. 445: 115086.

Sahadevan, R., K. Kannan, A. Jose, M. Porel and S. Sadhukhan. 2024. Unconventional fluorescence behaviour of ε-polylysine: Selective and sensitive detection of Cu^{2+} in real water samples, test paper strips, living cells, and anticounterfeiting applications. Sens. Actuators B Chem. 409: 135589.

Samanta, U., D. Pal and P. Chakrabarti. 2000. Environment of tryptophan side chains in proteins. Proteins. 38: 288–300.

Schiattarella, C., C. Diaferia, E. Gallo, B.D. Ventura, G. Morelli, L. Vitagliano, R. Velotta, et al. 2022. Solid-state optical properties of self-assembling amyloid-like peptides with different charged states at the terminal ends. Sci. Rep. 12: 759.

Sharpe, S., K. Simonetti., J. Yau and P. Walsh. 2011. Solid-State NMR characterization of autofluorescent fibrils formed by the elastin-derived peptide GVGVAGVG. Biomacromolecules. 12: 1546–1555.

Shukla, A., S Mukherjee, S. Sharma, V. Agrawal, K.V. Radha Kishan and P. Guptasarma. 2004. A novel UV laser-induced visible blue radiation from protein crystals and aggregates: scattering artifacts or fluorescence transitions of peptide electrons delocalized through hydrogen bonding? Arch. Biochem. Biophys. 428: 144–153.

Stephens, A.D., M.N Qaisrani, M.T. Ruggiero, G.D. Mirón, U.N. Morzan, M.C. González Lebrero, et al. 2021. Short hydrogen bonds enhance non-aromatic protein-related fluorescence. Proc. Natl. Acad. Sci. 118: e2020389118

Tomalia, D.A., B. Klajnert-Maculewicz, K.A.-M. Johnson, H.F. Brinkman, A. Janaszewska and D.M. Hedstrand. 2019. Non-traditional intrinsic luminescence: Inexplicable blue fluorescence observed for dendrimers, macromolecules and small molecular structures lacking traditional/conventional luminophores. Prog. Polym. Sci. 90: 35–117.

Trubetskaya, O.E., C. Richard and O.A Trubetskoj. 2016. High amounts of free aromatic amino acids in the protein-like fluorescence of water-dissolved organic matter. Environ. Chem. Lett. 14: 495–500.

Vijayan, V.N., K. Kannan, R. Sahadevan, A. Jose, M. Porel, S. Sadhukhan. 2024. ε-Poly-L-lysine: A naturally occurring biodegradable polypeptide for selective detection of 5-nitroimidazole antibiotics in animal products and living cells via fluorescence. ACS Appl. Bio Mater. 7: 4654–4663.

Weinryb. I. and F.R. Steiner 1970. Luminescence of the tryptophan and tyrosine residues of papain in solution. Biochemistry. 9: 135–146.

Wang, D. and T. Imae. 2004. Fluorescence emission from dendrimers and its pH dependence. J. Am. Chem. Soc. 126: 13204–13205.

Wang, D., T. Imae and M. Miki. 2007. Fluorescence emission from PAMAM and PPI dendrimers. J. Colloid Interface Sci. 306: 222–227.

Wang, R.B., W.Z. Yuan and X.Y. Zhu. 2015. Aggregation-induced emission of non-conjugated poly(amido amine)s: Discovering, luminescent mechanism understanding and bioapplication. Chinese J. Polym. Sci. 33: 680–687.

Wang, Y., X. Bin, X. Chen, Zheng, Y. Zhang and W.Z. Yuan. 2018. Emission and emissive mechanism of nonaromatic oxygen clusters. Macromol. Rapid Commun. 39: 1–6.

Wang, Q., X. Dou, X. Chen, Z. Zhao, S. Wang, Y. Wang, K. Sui, et al. 2019. Reevaluating protein photoluminescence: remarkable visible luminescence upon concentration and insight into the emission mechanism. Angew Chem. Int. Ed. Engl. 58: 12667–12673.

Xu, C., R. Guan, D. Cao, K. Liu, Q. Chen, Y. Ding, et al. 2020. Bioinspired non-aromatic compounds emitters displaying aggregation independent emission and recoverable photo-bleaching. Talanta. 206: 120232.

Xu, L., S. Zhong, Q. Meng, Z. Zhang, Y. Gao, X. Cui, et al. 2021. Novel material from natural resource, Agarose with clustering-triggered emission and its diverse applications. Dye. Pigment. 194: 109558.

Yan, J., B. Zheng, D. Pan, R. Yang, Y. Xu, L. Wang, et al. 2015 Unexpected fluorescence from polymers containing dithio/amino-succinimides. Polym. Chem. 6: 6133–6139.

Yang, G.Y., T. YeQiang, M. Ju, Z. Yiren, Y. WangZang, Z. YongMing, et al. 2013. Room temperature phosphorescence from natural products: Crystallization matters. Sci. China Chem. 56: 1178–1182

Zhang, H., Z. Zhao, P.R. McGonigal, R. Ye, S. Liu, J.W.Y. Lam, et al. 2020. Clusterization-triggered emission: Uncommon luminescence from common materials. Mater. Today. 32: 275–292.

Zhou, Q., B. Cao, C. Zhu, Y. Gong, W.Z. Yuan and Y. Zhang. 2016 Clustering-triggered emission of nonconjugated polyacrylonitrile. Small. 47: 6586–6592

Zhou Q., T. Yang, Z. Zhong, F. Kausar, Z. Wang, Y. Zhang, et al. 2020. A clustering-triggered emission strategy for tunable multicolor persistent phosphorescence. Chem. Sci. 11: 2926–2933.

Zuo, Y., Z. Gou, W. Quan and W. Lin. 2021. Silicon-assisted unconventional fluorescence from organosilicon materials. Coord. Chem. Rev. 438: 213887.

Index